Paradoxes of Western Energy Development

How Can We Maintain the Land and the People If We Develop?

AAAS Selected Symposia Series

 Published by Westview Press, Inc.
5500 Central Avenue, Boulder, Colorado

for the

 American Association for the Advancement of Science
1776 Massachusetts Ave., N.W., Washington, D.C.

Paradoxes of Western Energy Development

How Can We Maintain the Land and the People If We Develop?

*Edited by Cyrus M. McKell,
Donald G. Browne, Elinor C. Cruze,
William R. Freudenburg,
Richard L. Perrine, and Fred Roach*

Routledge
Taylor & Francis Group

LONDON AND NEW YORK

First published 1984 by Westview Press

Published 2019 by Routledge
52 Vanderbilt Avenue, New York, NY 10017
2 Park Square, Milton Park, Abingdon, Oxon OX14 4RN

Routledge is an imprint of the Taylor & Francis Group, an informa business

Library of Congress Catalog Card Number: 84-50794

ISBN 13: 978-0-367-28223-3 (hbk)
ISBN 13: 978-0-367-29769-5 (pbk)

About the Book

Proposed energy resource development in the arid western United States raises a number of potential problems for an environment that does not have a great deal of resiliency. Projected population increases associated with large-scale development activities may go beyond the capacity of small, isolated rural communities to absorb them; and constraints on western agricultural and industrial development--for example, demands for water already exceeding the supply available-- also limit energy development.

The authors of this wide-ranging book first evaluate western energy resources, then objectively discuss the consequences of development on the region's physical and social environments. Among the questions they consider are: Who will reap the economic benefits of development, and who will bear the environmental costs? What will be the effects on the environment? The social structure? The quality of life? Are open spaces a national treasure in their present form, or should they be regarded as space available for development? What are the unique demands of reclamation in the arid west? And, given the recent trend of western states-rights militancy and shifts of population to the southwest, what impact will new federal and state policies have on resource management?

About the Series

The *AAAS Selected Symposia Series* was begun in 1977 to
provide a means for more permanently recording and more
widely disseminating some of the valuable material which is
discussed at the AAAS Annual National Meetings. The volumes
in this *Series* are based on symposia held at the Meetings
which address topics of current and continuing significance,
both within and among the sciences, and in the areas in which
science and technology impact on public policy. The *Series*
format is designed to provide for rapid dissemination of
information, so the papers are not typeset but are reproduced
directly from the camera-copy submitted by the authors. The
papers are organized and edited by the symposium arrangers
who then become the editors of the various volumes. Most
papers published in this *Series* are original contributions
which have not been previously published, although in some
cases additional papers from other sources have been added
by an editor to provide a more comprehensive view of a
particular topic. Symposia may be reports of new research
or reviews of established work, particularly work of an
interdisciplinary nature, since the AAAS Annual Meetings
typically embrace the full range of the sciences and their
societal implications.

WILLIAM D. CAREY
Executive Officer
American Association for
the Advancement of Science

Contents

About the Editors and Authors

Cyrus M. McKell *is vice-president of research at NPI, a biotechnology firm in Salt Lake City, Utah. Botany, plant ecology, and natural resources management are among his specialties. He has written over 160 articles on arid land management, physiology of rangeland plants, land rehabilitation, land use planning, and shrub biology. He served on the Utah Council for Energy Conservation and Development and is currently chairman of the AAAS Committee on Arid Lands.*

Donald G. Browne *is a petroleum geologist working out of Denver, Colorado. His areas of interest include enhanced oil recovery techniques, comparative risk evaluations of energy systems, and the geology of the northern Rocky Mountains.*

Elinor C. Cruze *is senior associate at World Resources Institute in Washington, D.C. Trained in zoology, population biology, and ecology, she has written on sustainable resource development and energy use in agriculture. She is currently directing a policy study using ecosystem analysis to determine the consequences of major loss of biological diversity in tropical countries.*

William R. Freudenburg, *associate professor of rural sociology at Washington State University, Pullman, has specialized in social impact assessment and the policy-making process, especially societal decision-making on controversial issues. He has written more than two dozen scholarly papers on the social impacts of coal, oil shale, nuclear energy, and other types of energy development and is coeditor of* Public Reactions to Nuclear Power: Are There Critical Masses? *(with E. Rosa; AAAS Selected Symposium 93; Westview, 1984).*

Richard L. Perrine *is professor of engineering and applied science at the University of California, Los Angeles.*

He is the author of numerous publications on enhanced oil recovery; coal; the nuclear fuel cycle; solar, wind, geothermal, and biomass energy resources; and environmental and resource management. In 1975, he received the Outstanding Merit Award for Contributions to Environmental Engineering by the Institute for the Advancement of Engineering.

Fred Roach is an economist at Los Alamos National Laboratory. A specialist in resource and environmental economics, he has written more than twenty papers and reports on water and energy over the past five years.

David Abbey, a utility economist at Los Alamos National Laboratory, has written articles and reports concerning water use in the energy sector. He irrigates eight acres of pasture and orchard by the Rio Chama.

Stan L. Albrecht, professor of sociology at Brigham Young University in Provo, Utah, is interested in social impact assessment and social psychology research methods. A former editor of the journal Rural Sociology, he has done research on land use planning and the social impacts of Western energy development.

F. Lee Brown is professor of economics at the University of New Mexico, Albuquerque, and codirector of the Water Program for the Center for Natural Resource Studies in Berkeley, California. A specialist in water resources and econometrics, he is the author of The Southwest Under Stress: National Resource Development Issues in a Regional Setting (with A. V. Kneese; Johns Hopkins University Press, 1981).

Philip M. Burgess is professor of management and minerals policy at the Colorado School of Mines in Golden, Colorado, and director of the Institute for Minerals and Energy Management. From 1977-1982, he was president and director of the thirteen-state Western Governors' Policy Office and president of the WESTPO Institute for Policy Research. He has worked on energy projects involving international and industry-government cooperation in the Pacific Basin, North and South America, and Scandinavia.

M. Rupert Cutler is senior vice-president of the National Audubon Society in New York City. Formerly assistant professor of resource development at Michigan State University and Assistant Secretary for Conservation, Research, and Education, U.S. Department of Agriculture (1977-1983), he is involved in natural resource policy and management issues.

James H. Gary, professor of chemical engineering at the Colorado School of Mines in Golden, Colorado, is interested in fuels and refining processes. He is founder and director of the Annual Oil Shale Symposia, and his publications include Petroleum Refining, Technology, and Economics (with G. Handwerk; Marcel Dekker, 1975).

Gary B. Glass is a state geologist and director of the Geological Survey of Wyoming in Laramie. A specialist in coal geology, he has written more than sixty publications on coal in Wyoming.

Rita R. Hamm is a research associate for the Department of Rural Sociology and an energy information specialist at the Center for Energy and Mineral Resources at Texas A&M University. Her major area of research is the socioeconomic impacts of resource development.

John W. Hernandez is professor of civil engineering at New Mexico State University and a consultant in environmental management. Formerly deputy administrator for the Environmental Protection Agency, he has written extensively on water resources in the arid Southwest.

Peter W. House, currently director of the Division of Policy Research and Analysis at the National Science Foundation, has been involved in policy analysis studies related to land use, environmental resource management, and energy. His books include The Carrying Capacity of a Nation: Growth and the Quality of Life (with E. R. Williams) and Trading-Off Environment, Economics, and Energy (Lexington, 1976 and 1977, respectively).

Sean-Shong Hwang is a research scientist in the Department of Rural Sociology at Texas A&M University. A specialist in urban ecology and demography, he has done research on suburbanization and residential segregation.

Joseph G. Jorgensen, professor of comparative culture at the University of California, Irvine, is interested in comparative Native American cultures, past and present. He has done studies on the consequences of energy development on Native American tribes and on the adequacy of environmental impact statements. His books include Native Americans and Energy Development, I and II (Boston: Anthropology Resource Center, 1978 and 1983, respectively).

Allen V. Kneese is a senior fellow at Resources for the Future in Washington, D.C. He has written or contributed to numerous publications in resource economics, most recently Energy Development in the Southwest: Problems in Water, Fish, and Wildlife in the Upper Colorado Basin (edited with W. O. Spofford and A. T. Parker; RFF, 1980) and The Southwest Region

Under Stress: National Resource Development Issues in a Regional Setting *(with F. L. Brown, Johns Hopkins University Press, 1981)*.

Charles D. Kolstad, *an environmental and resource economist, is an assistant professor in the Department of Economics and the Institute for Environmental Studies at the University of Illinois. Formerly an economist at Los Alamos National Laboratory, he has been involved in studies of energy supply and international trade and the mathematics of applied economic equilibrium models.*

F. Larry Leistritz *is professor in the Department of Agricultural Economics at North Dakota State University, Fargo. Formerly a director of the Western Agricultural Economics Association, he is the author of* Socioeconomic Impact of Resource Development: Methods for Assessment *(Westview, 1981)*.

Arthur E. Lewis *is manager of the oil shale program at Lawrence Livermore National Laboratory. He has been active in oil shale research for the last ten years and has published many scholarly papers in this field. In 1982-1983, he spoke on oil shale as a distinguished lecturer for the Society of Petroleum Engineers.*

Steve H. Murdock *is associate professor and head of the Department of Rural Sociology at Texas A&M University. His specialties are demography and human ecology, and he has published widely on the demographic and social impacts of resource development. His books include* Energy Development in the Western United States: Impact on Rural Areas *(Praeger, 1979) and* Nuclear Waste: Socioeconomic Dimensions of Long-Term Storage *(Westview, 1983)*.

Richard B. Powers *is a geologist and project chief of the Western Thrust Belt for the U.S. Geological Survey in Denver, Colorado. He has worked in the field of petroleum geology exploration for more than thirty years in both North and South America and has published more than forty articles on oil and gas resource evaluation in the Western Thrust Belt and offshore U.S. basins. Among his publications is a two-volume edited book entitled* Geologic Studies of the Cordilleran Thrust Belt: Alaska to Mexico *(Denver: Rocky Mountain Association of Geologists, 1982)*.

James L. Regens *is assistant director for science policy in the Office of International Activities, Environmental Protection Agency. Currently on leave from the University of Georgia where he is associate professor and research fellow in the Institute of Natural Resources, he now serves* as *chair of the Energy and Environmental Group, Organization for Economic Cooperation and Development.*

William D. Schulze *is professor of economics at the University of Wyoming, Laramie. The author of more than sixty papers and reports, he has done his principal research on issues related to energy, including solar and geothermal, natural resources, and the environment. He has done statistical analyses of the health effects of air pollution and benefit analyses of reducing health risks from toxic substances. His other research interests include nuclear storage, water quality, earthquake hazards, and boomtown impacts.*

Michael D. Williams *is a staff member of the Energy Technologies Group at Los Alamos National Laboratory. An engineer and applied mathematician by training, he now does research on air quality issues and recently directed dispersion modeling for the National Commission on Air Quality.*

Introduction

Originally conceived as a symposium topic by the Committee on Arid Lands of the American Association for the Advancement of Science, the resulting series of papers highlights some of the major issues and paradoxes that must be addressed in developing western energy resources. Along with energy development, consideration must be given to the rights of western states to examine alternatives for commitment of valuable water, air, and space as well as how to deal with impacts and changes to their social systems that occur in energy resource development.

A treasurehouse of energy resources consisting of coal, oil shale, and uranium is scattered over the western states. These resources are needed by the nation to fuel industry, heat homes and factories, give motive power to the transportation system, and provide raw materials in the manufacture of industrial products.

Paradoxically, these valuable energy resources cannot be recovered and processed without committing some valuable and relatively scarce resources vital to the western states: water, air, and space. Water is available in the West but only at the price of shifting it away from present users. Air, or more appropriately, air quality, must be maintained in the vicinity of valuable national parks, recreation areas, and cities to meet established regulatory levels. Yet, it is possible to develop and utilize energy resources in appropriate areas. Maintaining appropriate levels of air quality is essential to the health of desert, valley, and mountain ecosystems in the west.

Other paradoxes exist in reconciling the need for jobs and opportunities for economic development while at the same time dealing with problems in community impacts, social change, urban sprawl, and adverse influences on native American cultures.

Federal and state government policies also face paradoxes in that they must consistently work for national energy self-sufficiency while at the same time enforce legislation requiring maintenance of environmental quality. Local and western region priorities and needs must be respected as well as addressing national energy and resource goals.

This series of discussions was presented at the AAAS meetings in Washington D.C. in 1982. Readers are now able to examine the facts and logic of professionals as they examine resources policy and legislation, development of energy resources, availability of air and water resources, and social issues regarding the paradoxes in energy resource development in the arid and semi-arid west. The book is organized in four sections:

(I)   Paradoxes in the implementation of environmental regulations and energy resource development policies, arranged and edited by Dr. Cyrus M. McKell and Dr. Elinor C. Cruze;

(II)   Western energy resources, arranged and edited by Dr. Richard Perrine and Dr. Donald G. Browne;

(III)   Water and air resources, arranged and edited by Dr. Fred Roach;

(IV)   Social and economic impacts of western energy development, arranged and edited by Dr. William Freudenberg.

Part 1

Paradoxes in the Implementation of Environmental Regulations and Energy Resource Development Policies

Cyrus M. McKell, Elinor C. Cruze

1. An Overview of the Issues

Proposed energy resource development in the arid west raises a number of major issues that bear on the question of what level of impact is acceptable in an environment that does not have a great deal of resiliency. Projected population increases associated with large-scale development may go beyond the capacity of small isolated rural communities to absorb them. Demand for water in the west exceeds the supply available, yet how can existing water supplies be reallocated equitably? Who will reap the economic benefits and who will bear the environmental costs of development? What effects will energy development have on the social structure, quality of life, and the arid environment? Are open spaces a national treasure in their present form, or should they be regarded as available space for development? What are the unique aspects of reclamation in the arid west that make rehabilitation so different from reclamation in other parts of the country? Given the new trend for western states'-rights militancy and shifts on population to the Southwest, what impact will new policies have on resource management? How should federal resource development policies provide opportunities for western states to make imputs into policy matters regarding the consequences of development on land, air, and water resources as well as the social conditions in the arid west?

Are federal regulations regarding environmental quality being implemented as vigorously and consistently as Congress intended or has there been a relaxation brought about by political pressure and budgetary constraint? All citizens have a right to expect that policies are fair and equitable for meeting regional as well as national needs and goals. Achievement of such a balance is difficult to attain in issues as conflicting and complex as associated with energy development in the fragile, social, economic and biological environments of the western United States.

2. Energy Development: A Western Perspective

Abstract

Mineral production is a driving force of the Rocky Mountain economy, and energy resources are the major component of that industry. With plentiful reserves of conventional oil and gas, plus coal, oil shale, tar sands, and uranium, the region already makes a significant contribution in meeting the nation's energy needs.

According to regional forecasts, investments over the next decade in energy projects and related community development could reach $50 billion. During that time, more than 200,000 new jobs could be created in the energy industry, and the region's population might grow by as much as one million persons. With the potential for such explosive growth--and some of it already under way--serious concern is voiced about the prospects for poorly managed "boom town" situations and the threat they pose to industrial developers and the regional economy. The pace of future energy development in the region, however, will depend on changes in current economic conditions (including world petroleum supplies and prices), the long-term future of the coal industry, government decisions regarding the synthetic fuels industry and mineral exploration in federal lands, and a host of environmental factors. Accordingly, sagging petroleum prices of other forces at work in the global or national environment could lead to an abrupt slow-down and to local communities holding the bag.

Introduction

It is my task[1] to provide a brief overview of the energy development activities in the western United States. Let me preface my views with some observations about why the Western Governors' Policy Office (WESTPO) is concerned about what is going on in the energy development area.[2]

The Rocky Mountain region is essential to the energy security of the United States because almost all the nation's new energy resources are located here. As we look to the next ten years, the Intermountain West will require a very substantial capital investment. In the energy area (synthetic fuels, oil and gas, coal and uranium) estimates range between $30 and $50 billion in that period. In addition to that, we can add something in the neighborhood of $10 billion for nonfuel mineral production.

Altogether, then, over the next ten years--not the next twenty or forty years--we can account for about $40 to $50 billion to be invested in the region for energy, nonfuel minerals, and for the supporting community and physical infrastructure. That is an unprecedented sum. To put in in perspective--$50 billion over ten years can be compared, for example, with the interstate highway program, the largest public works project in the history of mankind. But the interstate highway program that involved $78 billion was spread over 50 states, the District of Columbia, and a couple of the territories during a 30-year period. The manned Lunar space program cost $24 billion, but the money was spent in 43 states over 10 years.

When we look at the $50 billion that is already being invested in our part of the country, we must remember that this is a region with 41 percent of the nation's land area and only 5 percent of its population. This very substantial investment will impose tremendous requirements for materials and manpower in a section of the country that has the most fragile institutions to deal with those kinds of rapid changes.

The National Energy Situation

Let me examine what we see in the area of energy--and put it in a national context--and contrast our approach to estimating energy development with the approaches used by the Department of Energy (DOE), Exxon and others.

First, the national picture. I will refer to the Exxon version of the national picture because it has been in the news a lot, and because I think it has been misrepresented in many ways.[3] The Exxon view of the nation's other energy companies, of DOE, or of other forecasters. In fact, contrary to the implications of most editorials, the Exxon scenario is a fairly conservative one. Consider the energy demand forecast, for example: a 1.1 percent annual increase is estimated over the 1980's, followed by a 1.6 percent rate of growth from 1990 to 2000. By contrast, DOE's and other annual estimates go as high as 2.3 percent over the same period.

The Exxon plan is also conservative in a number of other senses: It assumes, for example, that we will be successful in streamlining government regulation. That remains to be seen. It also assumes there will be no interruption of foreign oil supplies. The whole business of the stability of foreign supplies is still a very "iffy" proposition.

If we examine energy demand in the United States by consuming sector, two important points emerge. First, in the residential-commercial area, substantial savings by conservation are plugged into this model; for example, in the year 2000 the Exxon plan assumes a 29 percent conservation factor compared with the years 1960-1973. This saving occurs in the residential-commercial consumption because the forecast envisions only a 1 percent increase in the total demand for that sector. There will actually be a decrease in transportation usage, reflecting the attrition of the pre-1973 cars from the fleet as well as the achievement of fuel efficiency of smaller cars and better engines. In the industrial area, while assuming substantial increases in output, only relatively small increases are expected to occur in fuel consumption. And in nonenergy areas, petrochemicals and others, demand remains constant.

On the supply side, according to the Exxon scenario, oil use will be tapering off, natural gas is beginning to stabilize, tremendous increases will occur in coal production--most of which, as indicated, will come from the western states. In addition to fossil fuels, nuclear energy is expected to grow from 5 to 13 percent of the total energy mix. This is another estimate that might be considered "conservative"-- in part because of the Three Mile Island incident and the fears it induced. Many people believe the nuclear future to be a lot less bright.

Liquid fuels are at the heart of the issue and have a real impact on the Intermountain West (the noncoastal western states). President Carter stated, in June, 1979, that our imports from abroad must never again exceed 1976 levels. That policy still stands and my guess is that it will be reaffirmed, if the occasion arises, by the Reagan Administration.[4]

As domestic supplies wane, and if we stick to our policy of never exceeding the 1976 bubble in imports, 1988 becomes a

very critical period. The late 1980's are dangerous because
our dependence on foreign oil will again reach a zenith.
According to the Exxon scenario, that gap will be filled by
synthetics, and the West will play a key role in that development. The transition ahead is one of declining oil and
gas supplies, nondepleting energy forms decades away, with
the 1980's and 1990's as the critical inbetween years for
conservation, nuclear, coal, oil and gas, and synthetic fuels.

The Essential Role
of the Rocky Mountain Region

Now I will focus on the West, because most of the new
energy resources are going to be developed in this part of
the country. First, we have oil and gas in both the traditional oil- and gas-producing area and in the Overthrust Belt.
In aggregate, these areas of the Intermountain West are now
producing 2.5 million barrels of oil a day, or 32 percent of
the total energy currently being produced in the United States
and 16 percent of its total energy consumption. In order to
put a frame around those figures--the nation's average consumption per day is 39 million barrels of oil equivalent,
about 15 million barrels of which are actually oil, and somewhat over half of those produced domestically. About 1.5
million barrels of the oil come from Prudhoe Bay, but the
Intermountain West minus Alaska still has a production level
of over a million barrels a day.

As for projections of oil and gas activity, very little
is accounted for by the Overthrust Belt, because this area is
only just opening up to intensive exploration. It now appears
that the Overthrust Belt, a formation that runs down the
spine of the Rocky Mountains, contains at least 1 billion
barrels of oil in recoverable, verified discoveries. (That
compares with Prudhoe Bay's 10 billion barrels.) Some estimates of the recoverable reserves in the Overthrust Belt run
as high as 15 billion barrels. This is a very substantial
potential for the Intermountain West.

Let me add a note about projections and how they are
constructed. When DOE and Exxon and others forecast energy
demand, they do it essentially from a deductive, analytical
modeling approach. Econometric methods, using carefully
worked out assumptions, generate estimates of future demand,
and aggregate amounts of fuel required, by type, to fill that
demand. Once the aggregate numbers emerge from the analysis,
others try to estimate where it will all come from and what
regions of the country are going to produce it.

WESTPO's effort supplements that approach starting from

an empirical, "bottoms-up" point of view. We go out to industry and ask what they have on line in all these areas.
Then we go to state governments and trace applications for
permits, or whatever. We are looking for paper trails that
exist in the private sector--in the banking community, in the
energy industry itself, and in the state government--that
will allow us to say that certain plants are going to go on
line in a certain county with a certain production and a
certain employment by a given year. With this approach, you
can look about seven to twelve years ahead--the average construction time for a typical energy facility. In our opinon,
this is a realistic approach for estimating the energy production likely to occur in 1985 or 1990.[5]

Abundant Resources

The Rocky Mountain region has, for all practical purposes,
100 percent of the nation's uranium. This is a very depres-
sed industry now--over a thousand workers were laid off at
the end of 1980 in New Mexico. It appears likely, however,
that the industry will be picking up by the mid-1980's. Our
estimates show that there will be about 23,000 new employees
in uranium by 1990--in existing uranium production (Fort St.
Vrain in Colorado and in Nebraska); in future uranium activity with new mines in Utah, Wyoming and New Mexico; a new
nuclear power station in Phoenix; and an expansion of the
Nebraska nuclear power station.

As for coal, the major producing areas are: Fort Union
in the Dakotas and Montana; the Powder River Basin further
South; the Hamsfork-Greenriver area, primarily in southwestern Wyoming; and the Unita Basin area in central Colorado
and Utah. Not all of the coal is surface mined in these areas,
although western mining is usually viewed that way in the
East. Underground mines, located mainly in Colorado and
Utah, produce coal that has a fairly high Btu content.
WESTPO has been making a vigorous effort to promote the export of western steam coal (for power generation) to the
Pacific Basin--particularly to Japan, Taiwan and Korea,
because of these countries' interest in high-quality coal.
Ultimately, much of the coal, however, is expected to go to
the synthetic fuels industry.

Altogether, western coal has made a tremendous contribution to domestic energy supplies. This region's coal production went from 6 percent of the nation's supply in 1970
to 26 percent in 1980 (173 million tons), and coal utilization is still on the rise. These figures do not square with
the "foot-dragging" image of the western states that is held
by many easterners. By 1990, we estimate that the western

states will be producing 432 million tons of coal, or just over half of total domestic production.

According to DOE figures, this region of the country will be producing 662 million tons in 1990, but WESTPO's estimates do not bear out that optimism--not because of environmental restraints, not because of community impact, but because when you go out of the industry and ask, "What's on line? where are the new mines going to open? how fast can you expand your present facilities?", the numbers that come back add up to 432 million tons. We realize that these production numbers change every year, but we want to begin to develop a dialogue between the states and the industry so that we can get better estimates of future production.

The West has also played a major role in the development of new power generation stations. More coal conversion plants have been built in the West than in any other region. Those power conversion stations are essential to the expansion of population and for the energy industry itself.

Oil shale development in the Rocky Mountain region is a major new development. For practical purposes, 100 percent of the recoverable shale is located in this region, essentially from the Piceance Creek Basin in Colorado and the Uinta Basin in Utah.

From our "bottoms-up" methodology, WESTPO has identified 36 synthetic fuel plants that are expected to be in operation by 1990. These break down roughly as follows: 19 coal liquefaction and gasification plants, principally in southern Wyoming and North Dakota; 12 oil shale installations in Colorado and Utah; 3 tar sands installations, all in Utah; and 2 ethanol plants with biomass supplied from agricultural projects in the region.

Altogether, then, based on known projects, the region can be expected to be producing about 900,000 barrels of synthetic fuel a day by 1990. That contrasts with DOE's target for the region of 2 million barrels in that year, and Exxon's target of 4 million barrels by the year 2000, of about 3 million barrels by 1990. The discrepancy lies in the methodology used to prepare the estimates. Bearing in mind that synthetic fuel plants take seven to twelve years to put in place, WESTPO's count includes only those for which permit applications have been initiated. Realistically, this means a much lower production level will be achieved in 1990 than projected by all other forecasts.[6]

Projected Employment

Over the next 10 years, the primary source of employment will be the synfuels industry; second, the oil and gas industries; third, coal; and fourth, uranium.

In addition to a general pressure on overall employment, there is, and will continue to be, severe competition for scarce critical skills, such as engineering and drafting. Over the next six years, for example, the demand for civil engineers will rise 50 percent--from 14,000 to 22,000. This heavy demand will undoubtedly result in a general inflation in wages which, in turn, will increase project costs for the synfuels industry and other energy development. A wage inflation in various industries will cause manpower shortages elsewhere in the region's economy.

Materials are also a potential problem. For example, if current plans for the MX missile deployment are implemented, there are serious questions as to how the requirements for cement and other materials will be filled. Some real constraints exist in both talent and materials in meeting the multi-billion dollar investments that are on the horizon.

Some states in the WESTPO region will be affected more than others. New Mexico, Utah, Colorado and Wyoming will bear the brunt of all aspects of energy development; Alaska will feel the push from oil; and synfuels activity will affect North Dakota and Montana. The other states in the region--South Dakota, Nebraska, Nevada, and Arizona--will be significantly less affected.

Projected employment growth in the synfuels industry will dominate future energy development in the region if first-generation technologies prove to be successful and are followed by a second generation. The oil and gas industry, the second most prominent energy employer, reflects the heavy development in the Prudhoe Bay area of Alaska. That activity accounts for somewhat over half of the employment in these industries.

Projected employment figures that have been described for the region are for direct, on-site employment only, and do not take into account secondary employment of population growth. Also unaccounted for is the potential development in the Overthrust Belt region, and any effects of Canada's evolving energy policy on Rocky Mountain development in the United States.

Other Factors Limiting Energy Development

Some additional constraints will affect the region's energy future. One of these is water. The Hundredth Meridian--cutting through the Dakotas, Nebraska, Kansas, Oklahoma and Texas--is where the West begins. It is also an important divider from an ecological standpoint: 73 percent of the rainfall in the United States falls to the east of that line; only 27 percent falls to the west--14 percent in the Pacific Northwest and 13 percent in the entire Intermountain West. The demands of energy development particularly, as well as the desire to maintain agriculture as a viable economic sector in the West, are going to put great pressure on scarce water supplies.

The transportation system in the West is another major problem. Our national transportation systems have been geared primarily to get people to California. But most of the coal produced must be transported by rail, and there are major problems with grade separations in rural areas where towns are divided by railroad crossings. And this is also true in cities such as Littleton (Colorado), where five unit trains, consisting of about 100 cars each, a day go through now and 15 unit trains are expected by 1985. WESTOP estimates that 222 grade separations will be needed in the Intermountain West just to handle the coal trains that will be passing through cities small and large.

These grade separations, by the way, raise an important public policy issue. Most of the coal trains in Colorado are merely passing through the state: Their points of origin, as well as their destinations, lie beyond the state's boundaries. Since Colorado is not involved in the revenue generated by this activity, questions arise as to the propriety and the equity of having Colorado taxpayers pay for the 59 railroad grade separations that are needed in that state.

Coal haul roads present another problem. Many need to be built. Many that were dirt roads several years ago now carry a sizable volume of traffic. The costs involved are enormous: In 1980 a joint federal-state effort estimated that about $20 billion will be required for the construction of coal-haul roads, particularly in the East. By comparison, the problem in the West is a relatively small one. About $406 million are needed in five states--Colorado, Montana, New Mexico, Utah and Wyoming. Altogether, over the next 10 years, the coal transportation system in this region alone will require $1.1 billion of investment: $406 million for coal-haul roads, $367 million for commuter highways, and $301 million for grade separations.

That, in essence, is the energy picture as we see it at WESTPO. Let me emphasize again that our estimates are not "spun out of the hat of an analyst." They are developed through the "grit work" of traveling around and talking to industry people, state government people, bankers and others, trying to verify the numbers on which expections are built. We intend to update these estimates as the situation continues to develop. And we also need to keep in mind that these are volatile numbers that reflect changing and uncertain times. A new SALT agreement, successful opposition to MX, an oil glut, or changing economic fortune with corresponding reductions in economic growth and energy consumption could change current resource development plans that are driving much of the economic development of the west. The base is broad. Activity is intense, but the roots are shallow and the commitments fragile, contingent, and hedged on all sides.

References

1. The author, now Director of the Institute for Minerals and Energy Management and Professor of Management and Minerals Policy, Colorado School of Mines, served as President and executive director of the Western Governors Policy Office (WESTPO) from 1977 - 1982.

2. WESTPO is an organization formed in 1977 by the governors of thirteen states--Alaska, Arizona, Colorado, Idaho, Montana, Nebraska, Nevada, New Mexico, North Dakota, South Dakota, Utah, Washington and Wyoming.

3. Exxon Company, USA's Energy Outlook, 1980-2000. Houston: December (1980).

4. The Reagan energy plan deemphasizes the role of imports and, although vulnerability is still a concern, there is also concern for too much restraint on the economy. In any case, high prices downsizing, and conservation efforts, as well as the current recession, have significantly reduced domestic demand.

5. For a detailed listing of energy projects in the Rocky Mountain region, see Ronald L. McMahon and Porter B. Bennett, Energy Activity in the West. Devner: ABT/west and WESTPO, March (1981).

6. The outlook for the synthetic fuels industry is even less optimistic today. In May, 1982, the Exxon-Tosco Colony Oil Shale Project--was cancelled. Major production companies appear to be experiencing new technologies, smaller scale projects, and joint venture opportunities that will reduce risk.

John W. Hernandez, James L. Regens

3. Environmental Regulations: The Western Energy Case

The Western Energy Potential

Environmental quality has become a national concern over the past decade. This is especially true in the Western United States where awareness has grown in direct proportion to the likelihood of large-scale regional energy development. The sharp increase in world oil prices over the past decade has made the West's enormous deposits of coal, oil, oil shale and uranium increasingly attractive (U.S. Department of Energy, 1980a:I-14). Energy development in the West poses a host of special challenges to accommodating accelerated production of energy resources while protecting the environmental integrity of some of the Nation's most scenic areas. Much of the West remains blessed with clean air and clean water, magnificent vistas, and untouched wilderness: a priceless national heritage. At the same time, the resources available for potential energy development are vast (Regens, 1978).

The U.S. Environmental Protection Agency (EPA) has completed a detailed technology assessment in order to identify and evaluate the environmental implications of Western energy development (White, et al., 1979). The study examined six energy resources (coal, oil, natural gas, oil shale, uranium, geothermal)[1] located in eight Northern Great Plains and Rocky Mountain States: Arizona, Colorado, Montana, New Mexico, North Dakota, South Dakota, Utah, and Wyoming. Table 1 reveals that this region contains 36 percent of known U.S. coal reserves, 6 percent of U.S. oil reserves, 8 percent of U.S. gas reserves, essentially all of the oil shale reserves, 90 percent of the Nation's uranium reserves and 22 percent of

---

[1]The study did not examine the following energy resources which also are in the region: hydropower, tar sands, solar and wind.

Table 1. Western energy reserve estimates.

Resources	Estimated[a] Reserves	Percent of U.S. Reserves from this Source
Coal	3,430	36
Oil	12	6
Natural Gas	20	8
Oil Shale	464	approx. 100
Uranium	264	90
Geothermal[b]	650	22

Q = quad (equal to 10^{15} Btu). One Q equals approximately 175 million barrels of oil, 60 million tons of western coal, or one trillion cubic feet of natural gas.

[a]Source: Devine, et al., (1981:3).

[b]This figure includes reserves, submarginal resources, and paramarginal resources.

our geothermal energy sources. Because this represents a substantial fraction of total U.S. energy reserves, a significant increase in the development of Western energy resources has the potential to change the environment and life styles of the region.

Mining activities and fuel-conversion facilities will generate solid waste. Construction and operation of synthetic fuel generating facilities and conventional power plants will consume water resources and will result in the release of some pollutants into the ambient environment. The manpower to construct and operate these mining and conversion facilities will produce substantial, rapid population increases in predominantly rural settings. New transportation systems are being and will continue to be developed throughout the region. The potential for social and economic conflicts is great but these problems can be mitigated by adequate and timely planning and financing of new community facilities (Federation of Rocky Mountain States, 1975; Devine, et al., 1981).

The U.S. EPA is committed to prudent development of the West's energy resources while maintaining environmental quality. Under a nominal national-demand scenario, energy production from the West would increase from 9 quads in 1978 to to 60 quads in the year 2000. Under a low-demand production scenario, energy production from the eight state study region

would increase from 9 quads in 1978 to 40 quads in 2000 (White,
et al., 1979). Western energy development can take place
with manageable environmental impact; EPA and the states involved share a responsibility to ensure that the adverse consequences of growth and development are acceptable.

Water Availability and Quality

Water resource management and conservation is one of several factors that will be important environmental considerations in Western energy resource development (U.S. Department
of the Interior, 1975; U.S. Water Resources Council, 1978;
U.S. Department of Energy, 1980b). The direct use of coal
for electric power generation consumes between 45 and 55
gallons of water per million BTU of electricity produced,
based upon the use of conventional wet-cooling towers. However, if dry-cooling towers are used, water consumption can
be lowered to 10 to 20 gallons per million BTU of electricity
produced. The same type of water savings can be achieved in
synthetic fuel manufacturing. For example, water consumption
for high BTU gasification can vary from 20 to 30 gallons per
million BTU of synthetic natural gas based on wet-cooling
towers to 10 to 18 per million BTU of synthetic natural gas
based upon dry-cooling towers (Devine, et al., 1981:61).
Clearly there will be a cost-differential, but this is where
trade-offs and management decisions must be made.

In the West, water quality is closely related to water
availability and could become an increasing concern as energy
resources are developed in the eight state area (Bishop, 1977).
By 1990, water requirements could exceed the minimum estimated supply (White, et al., 1979). Energy resource development in conjunction with increased water demands by other
users could result in the degradation of water quality (Grimshaw, et al., 1978; U.S. Department of Energy, 1980c). In
general, the value of water decreases with a decrease in quality to the point that some uses may be precluded. Two general responses can be adopted: improved effectiveness of
controls on water-use and control on pollution at the source
of generation. Ground water represents a large part of the
unallocated water resources of the region and its quality
must be protected.

Air Quality

Along with water, air quality will be greatly influenced
by Western energy development (White, et al., 1979). Conflicts between the goals of increasing energy development
and protecting the West's generally pristine air will directly affect which resources are developed, the technologies

used, and cost of the energy products. Most of the concern
over the impact of Western energy development on air quality
has focused on coal and oil shale. In general, the development and use of uranium, geothermal and oil and natural gas
resources will have less of an impact on air quality than will
either conventional or synthetic coal and oil shale applications (Devine, et al., 1981:113). Regional coal production
could become more and more important if economic conditions
remain favorable for export sales as well as growing domestic
markets (Western Coal Export Task Force, 1981). As a result,
several key EPA programs dealing with air quality regulation
will influence both the rate of Western energy development
and the environmental consequences of such development.

EPA's Prevention of Significant Deterioration (PSD) program has special importance for Western energy development
(Garvey, et al., 1978). All major new or modified sources of
pollution locating in areas where air quality is better than
the National Ambient Air Quality Standards (NAAQS) are subject to review under this program. This review ensures that
these sources apply "Best Available Control Technology"
(BACT) and that any resulting increases in ambient air concentrations do not exceed prescribed increments.

The current Clean Air Act and its 1977 Amendments (P.L.
95-95), establishes three classes of PSD areas. Class I includes national parks, monuments, and wilderness areas, many
of which are located in the West. Class II and III includes
all other areas of the country.[2] Due to the special nature
of Class I areas, the law allows only small increases in air
pollutant concentrations over these areas. The PSD regulations also require the protection of other "air quality related values" in the Class I areas, in particular protection
from visibility impairment.

By putting an emissions-cap on clean-air areas, the PSD
increment program has the potential to limit the amount of
industrial growth that can occur in these areas. Conceptually, once the increment has been used-up, no new-sources
would be allowed to locate in that airshed unless a corresponding reduction in pollution (i.e., offset) is achieved
for the area.

The PSD increment program can have another effect. To
verify that new sources of pollution will not violate a PSD
increment of the NAAQs, potential developers must provide an
analysis of air quality impacts, including estimates of the

---

[2]Only Class I and Class II areas have been designated.

effects that each project will have on soils, vegetation, and visibility in the area. This requirement has, on occasion, complicated the air emissions permitting-process and imposed sizeable costs on new projects (Arthur D. Little, 1980). For example, in one area where several energy-related facilities are scheduled for construction, the nondeterioration requirement for the collection of ambient air quality data prior to permitting has led to the operation of several different monitoring stations within a few miles of each other, all of them reporting much the same baseline data.

In the Clean Air Act Amendments of 1977, Congress directed EPA and the states to protect visibility in Class I areas. That legislation directed that decisions to retrofit controls on sources that impair visibility must be made by weighing the economic and energy costs of control against the benefits of the improvement in visibility. Because of the number and sensitivity of Class I areas in the West, the potential for visibility considerations to affect Western energy development is relatively high (Devine, et al., 1981:127-133).

The pollutants that cause visibility degradation are nitrogen oxides and very small-size particles. Gaseous NO_2 absorbs light at the blue-end of the spectrum and thus lends a reddish or brownish color to a plume. Nitrogen oxides are produced in fossil-fuel combustion and their greatest impact is close to the source of emission. Small particulates scatter light and create a haze that obscures visibility. Small particulates are not only emitted directly from pollution sources but may also be formed downwind as chemical transformation of gaseous primary emissions (primarily SO_2) occurs. Particulates formed at some distance from the source have time to disperse and to mix with pollutants from other sources. This, in turn, can make it difficult to identify the particular source from which the pollutant came. Source identification in the West is also complicated by natural sources of visibility impairment including windblown dust and forest fires.

Visibility regulations could result in constraints on siting of new facilities, but they would not necessarily require additional emissions control-hardware. Analyses performed so far indicate that the PSD increment, and not visibility, would be the limiting factor on energy development in the West (U.S. Department of Energy, 1980a:I-4). The primary means of achieving the PSD goals is through the use of permits. The PSD permit program is very complex and therefore, has the potential to cause delay, uncertainty, and additional costs for applicants.

Environmental Permits

From the late 1960s through the late 1970s, there occurred a number of cases where the progress of major Western energy projects was impacted adversely by the environmental permitting process. Among others, the list included: the SOHIO Pipeline project; the Kaiparowitz power plant; the Colstrip power plant; and the Intermountain Power Plant. In each of these cases, there are debates as to the effect of the permit process and its costs, but some delays and changes in decisions were probably directly related to the process.

Few major energy projects now seem to be delayed exlusively because of pending environmental permitting. What has happened? Have permitting problems been resolved? One explanation is that the permitting process has been improved (U.S. Department of Energy, 1980a:I-16). Table 2 reveals that other factors (i.e., equipment problems, labor disputes) are more frequent causes of project delays than are environmental regulations. Recently, a number of energy projects have been postponed or cancelled because of a deteriorating economy, a world-wide oil surplus, and increases in the natural gas supply as a result of deregulation. Thus, environmental permitting is not a major constraint to Western energy development in general but may be in specific instances.

Table 2.   Reasons for project delays for power plants construction between 1967 and 1976

Reason for Delay[a]	Number of Times Cited	Percent of Total
Vendor-related problems	154	37
Labor-related problems	142	34
Regulatory problems	51	12
Utility related problems	38	9
Legal challenges, weather, etc.	34	8
	419	

[a]It should be noted that most utilities gave more than one reason for delay for each plant.
Source: Western Governors Policy Office (WESTPO). Based on a survey of 150 utilities.

In order to ensure that the improvements in the permitting process are maintained as economic conditions improve, it is necessary to enhance its management so that the roles of the Federal government, states, industry, and public interest groups remain clearly delineated. Three broad initiatives in this area have been proposed in order to ensure that the permitting process addresses environmental considerations while allowing the market to determine the ultimate development of specific energy projects:

(1) Joint review process. A state team organizes a schedule and coordinates the permitting activities of all local, state, and Federal agencies which must grant permits for a major proposed new energy or industrial projects. This process has been working in Colorado for two years, and has recently been adopted by Utah, Illinois and Tennessee.

(2) Master permit processes. Some states believe the variety of individual permit requirements (air, water, solid waste, etc.) that a planned project must meet do not provide a comprehensive opportunity for state officials and the public to evaluate the desirability of the project. These states have established a master permit requirement, providing a forum for a "go" or "no-go" decision on the entire project. If the master permit is granted, all other necessary permits must be granted. If the master permit is denied, the project may not go forward. This master permit system clearly has not served to accelerate permitting, but it has focused and organized debate on major projects. California, Oregon, Washington, and Montana all have variants of this master permit system.

(3) Regulatory reform. Various measures to ease the burden on a developer seeking multiple permits have been suggested. These include: publishing directories of state permit requirements, assigning agency personnel to act as permit coordinates for projects, establishing deadlines for permit decisions by agencies, operating regional permit information centers, and so forth.

EPA is streamlining its existing administrative procedures. For example, in recognition of the importance of state-level involvement in air pollution control, the Clean Air Act incorporates a provision for State Implementation Plans (SIPs). In order to streamline the SIP revision process, EPA is encouraging parallel state and Agency processing

of SIP revisions. For non-controversial SIPs, EPA is encouraging conducting final rulemaking simultaneously with the
public comment period. The Agency also is encouraging limiting EPA review of non-controversial SIP revision to the appropriate regional office. These administrative changes are
intended to transform a complex and time-consuming process
into one in which the time and the capital-costs imposed by
the process are commensurate with its social benefits.

Western Coal Production

New coal-fired power plants are now required to meet
both an emissions limitation and an overall percentage reduction of sulfur from the coal they burn. This percent reduction provision generally requires costly treatment-tech-
nologies even if the emissions limit could be met by burning
low-sulfur coal (U.S. Environmental Protection Agency, 1979).
Under the percent removal standard, consumers in the 1990s
will pay about $2-3 billion more annually than they would
have paid if the 1971 uniform standard were still in effect.
Flue gas scrubbers add about 5-20 percent ot the capital cost
of a new power plant, or about $25-100 million on a typical
500-megawatt boiler (Electric Power Research Institute, 1979).
Over half of these additional costs will be borne by electricity consumers in the West (ICF Inc., 1981).

The percentage reduction requirement was based partly on
concern that Eastern and Midwestern utilities would meet
emission limits by using low-sulfur Western coal rather than
locally available high-sulfur coal, causing market and employment dislocations (Ackerman and Hussler, 1981). It is now
apparent that transportation costs for low-sulfur Western
coal often make that alternative somewhat less attractive for
those regions than the use of Eastern coal. Removing the
requirement would have little or no effect on current mining
operations but there may be some post-1990 regional shifts in
coal production. These shifts will be minimal because of the
rapidly increasing costs of coal transportation. Moreover,
these changes are small relative to the substantial increases
in production projected to occur in most coal producing regions over the next twenty years. EPA also projects that
switching to a uniform standard would increase Mountain and
Pacific region sulfur dioxide emissions by no more than
200,000 tons in the year 2000 (ICF Inc., 1981).[3]

---

[3]This is a relatively small increase since smelters now emit
1.2 million tons of SO_2 annually in those two regions.

The allowable emissions of SO_2, NO_x, and TSP in coal combustion have been significantly reduced since 1971. At that
time, emissions up to 1.2 pounds of SO_2, 0.7 pounds of NO_x,
and 0.1 pounds of particulate matter were permitted per
million BTU of coal fired in new utility boilers. In most
cases, this represents a 50 percent reduction in allowable
air emissions from new coal-fired utility boilers.

Significant advances have been made in flue-gas desulfurization processes. In the future, EPA's dry SO_2 control program may yield significant improvement in environmental controls. The dry-process, LIMB technology (limestone injection
with multi-stage burners) has been shown to remove sulfur
oxides on brown coals in Europe (Hein and Glaser, 1980), and
the results to date for bituminous coal appear encouraging.

Synfuels

In cooperation with state environmental offices, EPA is
working to insure the environmentally safe commercialization
of the synthetic fuels industry as it begins to grow. The
progress of the synfuels industry is quite uncertain at the
present time because of the changing economics of natural
gas and world-wide oil production.

Technology exists to limit air emissions of sulfur and
nitrogen compounds in synfuel manufacturing to levels which
are lower than an equivalently sized, coal-fired thermal
power plant. In fact, steam generation at a synfuel plant is
the largest single source of sulfur oxides in most processes
using coal. While no full-scale synfuel plant exists in the
United States, EPA has obtained data for environmental
assessments and control technology evaluations from foreign
facilities (U.S. Environmental Protection Agency, 1981).
Analyses based on these data support the conclusion that a
synthetic fuel facility can be constructed in the United
States that would not result in any unacceptable environmental degradation.

Summary

The emphasis of the last decade on regulatory controls to
protect environmental quality has led to uncertain, changing
and occasionally ambiguous pollution control requirements,
redundant and overlapping state and Federal roles, and

---

[4]The emissions limit is 1.2 pounds of SO_2 for Eastern high-
sulfur coals.

expensive, complex, and time-consuming administrative processes. We should seek to reduce the time, costs, and uncertainties involved in different regulatory processes, while
at the same time achieving the maximum effect from every
dollar spent on pollution control technology.

A key objective underlying any regulatory reform is the
establishment of an adequate scientific and technical basis
to support a cost-effective approach to a cleaner environment. The uncertainties implicit in any large-scale economic
endeavor such as Western energy development must not be com-
pounded by uncertainties, either in Federal environmental regulations or in their scientific bases.

We do not believe that environmental legislation pre-
cludes substantial energy growth in the West. Instead, a
balanced approach relying upon clear regulatory signals to
complement the market mechanism, can ensure that the United
States achieves both energy growth and environmental quality
in the West.

References

Ackerman, B.A. and W.T. Hussler. Clean Coal, Dirty Air. (New
Haven: Yale University Press (1981).

Bishop, B.A. "Impact of Energy Development on Colorado River
Water Quality". National Resources Journal 17, October.
(1977).

Devine, M.D. et al. Energy From the West. Norman: University of Oklahoma Press (1981).

Electric Power Research Institute. Technical Assessment
Guide. Palo Alto: EPRI (1979).

Federation of Rocky Mountain States. Energy Development in
the Rocky Mountain Region: Goals and Concerns. Denver:
Federation of Rocky Mountain States (1975).

Garvey, D.B. et al. The Prevention of Significant Deterioration: Implications for Energy and Development. Argonne,
IL:Argonne National Laboratory (1978).

Grimshaw, T.W. et al. Surface-Water and Ground-Water Impacts
of Selected Energy Development Operations in Eight Western
States. Austin, TX:Radion Corporation (1978).

Hein, K. and W. Glaser. "Dry additive process for SO_2 removal during combustion of brown coals," presented at 6th

Member Conference, International Flame Research Institue (1980).

ICF, Incorporated. Analysis of Alternative NSPS Regulations-draft report prepared for U.S. Environmental Protection Agency, Energy Policy Division. Washington, D.C.: ICF. October, (1981).

Little, A.D., Inc. The Effects of Prevention of Significant Deterioration on Industrial Development. Cambridge, MA: Arthur D. Little, Inc. (1980).

Regens, J.L., Inc. "Energy Development, Environmental Protection and Public Policy." American Behavioral Scientist 22 November/December, (1978).

U.S. Department of Energy. Energy and Environment: Issues, Information and Constraints. Washington: Office of Environmental Assessments. July, (1980a).

U.S. Department of Energy. Institutional Constraints on Alternatives Water for Energy. DOE/EV/10180-01. Washington: Office of Environmental Assessment. November (1980b).

U.S. Department of Energy. Ground Water and Energy. CONF-800137. Washington: Office of Environmental Assessments. November (1980c).

U.S. Environmental Protection Agency. Environmental Assessment: Source Test and Evaluation Report (Lurgi-Kosovo final report). EPA-600/7-81-142. Washington: EPA. (1981)

U.S. Department of Interior. Westwide Study Report on Critical Water Problems Facing the Eleven Western States. Washington: Government Printing Office. (1975).

U.S. Water Resources Council. The Nation's Water Resources: 1975-2000, Second National Water Assessment. Washington: Government Printing Office. (1978).

Western Coal Export Task Force. Western U.S. Steam Coal Exports to the Pacific Basin. Denver: Western Governor's Policy Office. (1981).

White, I.L. et al. Energy From the West. Washington: U.S. Environmental Protection Agency. (1979).

4. What Has Happened to Enforcement of Environmental Regulations?

Environmental regulations do not exist in a vacuum. Rather, they interpret, and aid in the enforcement of, the acts or statutes passed by legislative bodies. They are presumed, therefore, to reflect the desires of the legislator's constituents. . .the voters. . .the taxpayers.

Enforcement is carried out by human beings. Their social political and economic orientation influences how they interpret their job responsibilities. . .how they exercise their administrative discretion.

Budgets determine the size of inspection and enforcement staffs, their ability to travel, and the backup available to them through, for example, agency solicitors' offices and the Department of Justice.

So we must touch on all these areas--statutes, staffing, and spending--if we are to answer comprehensively the assigned question, What has happened to the enforcement of environmental regulations?

The role of government in this context is to sponsor a balanced mix of environmental protection programs and economic development opportunities, for, in the last analysis, our lives depend on our natural environment--on air, water and the soil on which we grow crops--at least as much as we require jobs, factories, transportation, raw materials, and other components of our complex economic system for our own livelihood.

So I shall take the liberty of rephrasing the question put to me, and ask: Are the Nation's environmental regulations, as currently being interpreted, re-written and enforced, directed toward providing this healthy balance between the needs of the environment and the needs of the economy?

In Audubon's judgment, the answer is no. The current federal effort is off-center, out of balance, biased in favor of commercial developmnet, weak with regard to achieving environmental protection objectives. Enforcement of environmental regulations by the federal government recently has been handicapped, half-hearted, even occasionally perverse in its environmental consequences.

I'll be happy to explain this negative conclusion.

Most of the laws we are concerned with here today were passed within the last 20 years. They came into existence to ameliorate the very real adverse effects of unprecedented new pressures on the natural environment created by an expanding human population and its demand for energy.

There were two billion people in the world when I was born. Today there are more than four and one half billion. And they use more energy per person. More factories, more power plants, more automobiles brought more air pollution. Bigger draglines and other mechanization brought deeper coal mines with attendant water pollution. New synthetic chemicals spread through the planet's biosphere; PCB and DDT were found in the tissues of penguins in the Antarctic. Land-wasting suburban sprawl began to result in congested megalopolises like the Northeast Corridor, the Colorado Front Range, and Southern California. These sorts of problems touched the lives of large numbers of Americans. Citizens sought relief. Sponsors were found, and laws were passed, often with few dissenting votes.

While there is no need to belabor the point--this history is well known--it is important, as we discuss enforcement of environmental protection laws, to remember why these laws were enacted and the context in which they were framed. I have in mind particularly the laws to control pollution such as the Clean Air Act, the Clean Water Act, and Resource Conservation and Recovery Act, the Toxic Substance Control Act with its Superfund authority to clean up abandoned chemical waste dumps, and the National Environmental Policy Act, all passed during the years that have come to be known as the Environmental Decade. I am speaking as well of the laws passed to promote orderly planning for use of the federal public lands; these include the Multiple-Use, Sustained-Yield Act, the Forest and Rangeland Renewable Resources Planning Act, the National Forest Management Act, and the Federal Land Policy and Management Act. And I am speaking of the new systems created to protect the fast-vanishing remnants of our natural wildlife/wildlands heritage: the Endangered Species Act, the Wilderness Act, the Wild Rivers Act, the Land and

Water Conservation Fund Act, the Alaska Lands Acts and all the laws that have been passed to place additional lands and waters under the protection of the wilderness and wild river systems.

Let us remember, then, that the laws behind the regulations we are discussing were passed in response to major new problems that were troubling the Nation, that they were developed through years of study and public hearings, and that they have had the strong active support of past Republican as well as Democratic administrations.

Let us look at the enforcement of a particular statute, the Surface Mining Control and Reclamation Act of 1977, as a case in point:

I have a clipping from the August 4, 1977 issue of The New York Times reporting on the enactment of that law. It is headlined, "President Signs Strip-Mining Bill; Climax of Ten-Year Fight Marked by Ceremony." It was a front page story; the headline where it continued on an inside page read: "President Signs a 'Watered-Down' Strip-Mining Bill." 'Watered-down' was a quote from President Carter, who said, "I would have preferred a stricter bill."

Conservationists considered it a major step forward. After years of widespread failure by state governments to control surface mining practices and require reclamation, a start was being made toward workable, uniform federal standards to protect people, property and the natural environment. Consider what is happening to enforcement of that landmark law today:

•Although the Congress refused to make all the deep cuts asked for by the Reagan Administration in funding for inspection and enforcement by the Interior Department's Office of Surface Mining, still these activities have been decreased by approximately one-half.

•A new policy allows mine operators to leave "high walls" in violation of the law's requirement to return strip-mined lands to their approximate original contour. The operator must merely pay a minimal penalty, called by critics "paying for your high wall".

•Another recent OSM policy change, providing for so-called "compliance visits", reduces opportunities for third-party enforcement. It encourages federal and state inspectors to negotiate with mine operators in correcting violations, rather than to issue cessation orders or notices of of violation as required by law.

•Secretary Watt recently issued a regulation that would allow many existing strip mines on prime farmland to side-step the legal requirement that mined land be restored to productive condition. An environmentalist lawsuit has been filed to challenge the Secretary's omission of a deadline from a regulation he issued in September to implement the so-called grandfather clause of the Surface Mining Act which established exemptions for mines operating at the time the law was passed. The Carter Administration's regulation provides no practical limits to strip mining on prime farmlands. Remember: we're losing one million acres of prime farmland to urban conversion every year.

•There has been little effort to act against major coal companies that have been mining coal through subcontractors in Virginia and other southern Appalachian states. This ruse has enabled big companies to qualify for exemptions permitted at mining operations of less than two acres.

•Under proposed revisions of OSM regulations, citizens will no longer be assured of access to mines, they will no longer be allowed to bring citizen suits, they will no longer be able to participate in administrative proceedings, and they could not be awarded attorney's fees.

•Under the budget proposed by the Reagan Administration for the current fiscal year--a proposal the Congress refused to accept--there would have been less than 70 federal inspectors enforcing the Surface Mining Act nationwide. As it is, 66 inspectors and 26 support positions are being phased out in Fiscal 1981. The proposed budget for Fiscal 1982 assumes that all 24 states which submitted their own state programs for approval before the deadline of March 3, 1980, will have implemented their approved programs. Therefore, the budget projects further staff reductions of 87 inspectors, 80 support personnel, and 22 executive direction and administrative personnel by the end of Fiscal 1982.

•An associate solicitor in the Interior's Solicitor's Office, in its Division of Surface Mining, returned a number of proposed injunction cases involving violations of various performance standards to a field office with orders to determine how much it would cost the companies to abate the violations, and how much harm would be done to the public and the environment if the violations were not corrected. He asked his field staff to identify violations where the cost of correction was high and the damage low, and to give them low enforcement priority.

The adverse social and environmental effects of uncontrolled strip mining have been widely discussed over many years. They are well known, and there is no need to review them here. The 1977 law is far from being fully in effect. The social and environmental abuses it was intended to ameliorate--ugly, dangerous, unproductive, pollution-producing mined lands and sick communities--still are very much with us. It looks as though they may be with us far longer than the law's sponsors expected would be the case.

Another relevant clipping from The New York Times is datelined Harlan, Kentucky, December 18, 1981. It reads as follows:

"One woman was killed and more than 100 fled into snow and 20-degree weather today after a wall of thick muck from a coalfield sludge pond rumbled down on the small community of Ages.

The mixture of mud, mine waste and timber debris covered a mile-long area and spilled into a highway leading into Harlan, five miles to the east. In spots it was ten feet deep.

At least three homes were demolished and 15 other were damaged.... The body of Nellie Woolum, 65, a retired postmistress, was found in the ruins of her home.

"The disposal area was examined Monday and declared safe by an inspector from the Harlan office of Mine Safety and Health Administration.

"Governor John Y. Brown declared a state of emergency."

End of news story.

Interior's associate solicitor for the Office of Surface Mining proposes to weigh the cost of correcting a violation against the risk of damage from leaving the violation uncorrected. The implication of his proposal, and of other changes now being made in environmental regulations, is that the dollar cost of compliance with the law easily can be weighed against the dollar benefits to the public, and that "uneconomic" compliance will not be required. In fact, it is more than an implication. Administration-supported legislation for regulatory reform now pending in the Congress (regulatory reform being the euphemism used for eliminating regulations and procedures that hinder industry), would require that the costs involved in complying with federal regulations be shown to be less than the dollar value of the expected results.

The final arbiter of this cost/benefit assessment process would be the President's Office of Management and Budget.

I have seen that Office at work under political pressure, as when the 1980 Forest Service Resources Planning Act Program recommendations produced by USDA's professionals were drastically changed in response to White House direction. And public confidence in the OMB operation was not exactly enhanced by the admission of its current Director, David Stockman, that "None of us really understands what's going on with all these [budget] numbers" and his reference to "the greed that came to the forefront" during White House bargaining to pass tax-cut legislation, as reported by William Greider in the December 1981 Atlantic Monthly.

Conservationists doubt that this OMB--or any other agency--is competent to tote up the benefits of preventing a mud slide. While the value of the destroyed and damaged homes can be estimated in dollars and cents, the costs of the inconvenience and discomfort suffered by those who had to flee their homes, the uninhabitable and damaged environment, other short and long range effects on the little community, and of course, the value of the life of the retired postmistress cannot easily be assigned a cash figure.

Obviously, there is a limit to what we can pay for environmental protection. A proposed regulation that puts a heavy burden on many people to achieve small benefits for a few cannot be justified. The National Audubon Society does not pretend to know where the line should be drawn. We do say it involves intangibles that cannot neatly be fitted into a mathematical formula at the OSM or by the OMB. Such balances can only be struck in the political process, not by a mathematical process...thrashed out through open debate and voting in legislatures. Then the targets established by the legislative branch should be actively pursued by the executive branch through strict enforcement.

The direct effects of pollution on health often cannot be measured with certainty. An Audubon scientist who offered testimony recently in a case involving a power company quoted statistical studies indicating that if the scrubbers were not installed, the sulphur emissions would cause the deaths of between 2 and 150 people in the greater New York City area per year. The median projection was 15 deaths per year. He conceded it was not possible to prove the emissions would would cause any deaths, drawing an analogy with the relationship between smoking and lung cancer. "It was never really possible to prove that cigarette smoking caused cancer," he said, "yet few now doubt that cigarette smoking is deleterious to health".

If statistical data of this nature can be inconclusive, how can the OMB be expected to apply cost/benefit analysis to the cost of scrubbers as against the annoyance of smarting eyes, or a head cold, or various cumulative effects that might, or might not, be attributed to air pollution? Our contention is that these are not purely dollars and cents problems. Environmental regulations geared principally to accommodate the short-term cash-flow or balance sheet problems of polluting industries are out of balance with the Nation's long-term need to protect its life-support system.

To stress the need for balance between environmental and economic considerations does not assume that the two should be seen as competing goals. Quite the contrary. Both economic and environmental health are essential and obtainable. Let me pursue that point through brief discussion of the wilderness program and policy of the USDA Forest Service.

Passage of the Wilderness Act in 1964 assured that carefully selected roadless areas, truly qualified for inclusion in a National Wilderness Preservation System, would be set aside to be protected in their wild state for all time. An inconsistent section of the act permits--but does not require--the executive branch to issue mineral leases in existing wilderness areas until midnight, December 31, 1983. All previous Republican and Democratic administrations have directed the Forest Service to turn down applications to dig and drill in wilderness areas, because of the obvious conflict with the principal purpose of the wilderness designation. The discretionary power to issue mineral leases has never been used by any previous administration.

However, the current Assistant Secretary of Agriculture for Natural Resources, John B. Crowell, Jr., has adopted a different policy. It appears to be designed to <u>encourage</u> oil and gas companies to apply for leases in wilderness areas. Forest Service regulations formerly stated the service "would not normally recommend approval" of leases in wilderness. This regulation was changed in 1981 to read, "Designation of an area as wilderness may not be the basis for denying a mineral lease." The first such leases were granted last September for lands within the Capitan Mountain Wilderness in New Mexico. This wilderness mineral leasing--without an environmental impact statement--caused quite a negative reaction in the Congress--a reaction that has yet to be fully resolved, and which could result in amendatory legislation to cut off mining in wilderness even earlier than 1984.

There are many reasons why the government, through the democratic process, has decided to set aside wilderness areas for special protection. One reason was admirably expressed by Governor Bruce Babbit of Arizona, in a recent issue of Arizona Highways. Said the governor, "Wilderness guarantees our children will have a chance to experience the raw flavor of the Old West as a guide to knowing and understanding the New West they will inherit in their time."

Wilderness has many public uses. Watershed protection is a vital one. Many people forget that the National Forests were created just as much to protect watersheds as to provide timber. Forest Service lands produce as runoff about three-quarters of the West's water supply. There is no resource of greater economic value to the West than water, and no better way to assure its continued flow and high quality through wilderness designations. Wilderness provides non-motorized recreation of many kinds, scenic preservation, the conservation of diverse genetic material including many species of wildlife that require a wilderness habitat, and opportunities for teaching and research. In other words, wilderness provides for multiple use of a non-commercial kind.

We have all heard the charge that wilderness is a lock-up of natural resources which ought to be opened to other uses. The truth is that gas, oil and mineral development are single uses--and that they effectively exclude the public. The wilderness mineral leases being promoted by the current administration represent a private use of public lands. The charge of lock-up doesn't wash.

National Wilderness Preservation System components were selected only after lengthy hearings at which all possible uses and values were examined. All interested parties were involved in drawing the lines on the map. As one result, today's wilderness system is estimated by a Wilderness Society consultant to overlie no more than one percent of the Nation's potentially producible gas and 1.4 percent of its potentially producible oil. In fact, the Congress and agencies involved have worked hard to minimize the opportunity cost of wilderness designations. Conservationists often claim the wilderness system gets only the lands no one else wants. "Wilderness on the rocks," they call it. Meanwhile, demand for energy has declined while estimates of U.S. energy reserves-- outside of wilderness areas--has gone up. There is no emergency need to exploit these energy resources. Similarly, there is little basis for the charge that important volumes of commercial timber are being locked up in wilderness. Industry already has contracts for standing timber in the National Forests amounting to some 33 billion board feet--the largest backlog of standing timber under contract in the history of National Forest timber sales.

The wilderness system covers some 80 million acres, but most of it is in arctic Alaska. It covers about 2.4 percent of the total acreage of the United States, but only about one percent of the lower 48 states. One percent of the nation's commercial forest land in found in wilderness, but providing road access to this timber would be difficult and expensive.

Wilderness should be the last place we cut, dig or drill. The Nation still is far from any emergency which would justify scraping the bottom of its resource barrel by invading its wilderness areas.

One environmental statute that has come in for some ridicule is the Endangered Species Act, the law under which an obscure species of little fish held up construction of a multi-million-dollar dam. It was reauthorized in 1982, but came under attack from those who considered it a frivolous block in the path of important hydroelectric power and other construction projects.

There are scientific, practical reasons for protecting plants and animals from needless extinction. The rate at which species are disappearing is increasing, as expanding human settlements level rain forests, fill wetlands, and destroy other specialized habitats on a grand scale. As many as a million species--about one-fourth of the diversity of life on earth--may become extinct within the next 30 years. Virtually all of mankind's food and much of its fibers, medicines, and chemicals come from, or are derived from, natural species. No one knows what potential benefits may be lost to mankind if those million species are lost before science can study them. Endangered species protection is no joke. Yet, the budget for the federal endangered species program recently has been cut by 40 percent.

And drastic cuts also have been announced or are under consideration for the Environemntal Protection Agency, the principal coordinator and enforcer of the Nation's environmental protection laws. Overall, the EPA's budget currently is down 20 to 25 percent in both personnel and dollars. There is an additional "deferral" cut of 12 percent being worked out for 1982, and a 12 to 20 percent cut reported to be planned for 1982. This comes to a one-third cut in staff--4,000 fewer employees--and a 50 percent cut in budget over the next two years for the agency responsible for environmental protection across the board, from air and water pollution control to pesticide regulation--and the government's newly undertaken war on toxic substances and hazardous wastes. Staff morale at the EPA is shot; its professionals are leaving in droves.

To look at these budget and staffing cuts from another point of view, here's what the new budget has been bringing in terms of judicial remedies. In 1980, the EPA referred 230 cases to the Justice Department, and 180 of the cases were filed. In 1981--up to the time of the computer printout at the end of September--the respective figures were 65 and 53. Moreover, only 15 of those cases were referred and only a dozen filed after the present EPA adiminstrator took office, last May. If the EPA continues at that rate of referrals, it will be referring only 29 cases a year, as compared with from 150 to more than 200 in years past.

It is the size of the environmental enforcement budget cuts that troubles us. The Reagan Administration was elected to office on the basis of a compaign that pledged to cut government spending. Conservationists cannot object in good conscience to "their" programs taking a fair share of the brunt of government economies, nor can they object to cuts that eliminate waste and inefficiency. We do contend, however, that the pattern of disproportionately heavy cuts in environmental programs indicates the Reagan Administration believes environmental health is a frill the nation can't afford during a period of economic hard times.

Important examples in this regard are energy conservation and renewable energy programs. The Department of Energy, which itself appears to be on the way to oblivion, is cutting back its solar research program by 60 percent and its program for encouraging energy conservation by 75 percent, while increasing subsidies for nuclear power and providing encouragement of oil development.

Energy resource development often creates environmental problems. Energy policy cannot ignore pollution effects and regulation needs. The way the human race has produced, transported and consumed fuel--both fossil fuel and wood--during the past two centuries has had a devastating effect on the natural environment and on the health of plants and animals including people. The principle pollutants that threaten our health today are carbon monoxide, sulfur oxides, nitrogen oxides, acid precipitation, hydrocarbons, particulates, coal dust, nuclear wastes, and toxic chemicals. All but that last one come from production and use of energy.

It follows that an important way to protect the environment--and human health--is to minimize energy use. The federal government currently subscribes to the opposite policy.

In the belief that more energy use would be good for the economy, it aims at increased fuel production.

Two years ago, the National Audubon Society began an expansion of its science division by taking on its first major environmental policy study--on energy. We collected facts and figures from all available sources, including oil companies, government agencies and university research teams. We estimated the costs of the alternatives, and came to the following conclusion:

A combination of energy conservation and solar power would cost far less in dollars, promote a healthier economy, do more to strengthen national security, and do much less damage to the environment, than an all-out effort to produce more oil, nuclear power, and synthetic fuels. When we talk about energy conservation, we do not mean lowering our standard of living. We are talking about using available energy more efficiently. We contend that, by shifting to technologies that are more efficient, this country could be producing 50 to 80 percent more goods and services in the year 2000 without any increase over present energy consumption.

At the outset, I observed that enforcement of environmental regulations is influenced by the people who enforce them. Many of the top posts in the federal agencies charged with enforcing environmental regulations now are filled by men and women who have publicly declared they believe federal environmental regulations have been too strict, are a "drain on capital," and are harming the economy. They have set out to change the balance. Are they truly correcting an imbalance, or are they overreacting? Have they the expertise and understanding to find the optimum balance between encouraging economic development and protecting the long-term productivity of the land and the health of the people?

Certainly, no official's performance in a new assignment should be prejudged by his or her earlier views and background. But, I suggest the cumulative weight of the records and public statements of the current appointees directing federal environmental agencies indicates that a one-sided, pro-development view predominates.

I am not faulting these people as individuals. The fact that a particular government official's view displease the National Audubon Society is hardly a condemnation of an entire administration. Audubon recognizes that there is need for improvement and simplification of the laws and regulations governing the Nation's newly undertaken environmental protection program. Having conceded this, we suggest that the preponderance of the evidence--the views and backgrounds of the people in charge of the programs; the performance record to date, including the vacating of suspension orders and notices of violations--point to a decided lack of vigor in the enforcement of environmental regulations.

The federal goal to reduce U.S. dependence on foreign oil involves increasing coal mining from the almost 830 million tons mined in 1980 to well over one billion tons by 1990. If fragile western land and water resources are to be preserved, essential and achievable reclamation and water quality goals should not be set aside.

Our wilderness, our watersheds, our life support system, and our health should be forcefully protected by government.

Yet, our national government's efforts to enforce mine-reclamation and other environmental protection regulations currently are aimed more at easing the regulatory burden on business and industry than at following the policy direction the Congress provided in the environmental statutes. They fall short of the mark to the extent that they fail to recognize we cannot have a healthy economy without a healthy environment and a healthy citizenry to support it and to participate in it.

And _that_ is what has happened to enforcement of environmental regulations.

Part 2

Western Energy Resources

_____ *Richard L. Perrine, Donald G. Browne*

5. An Overview of the Nation's Treasurehouse

In his intriguing book, The Nine Nations of North America (1), Joel Garreau calls the portion of the United States that is the region of concern for this symposium "The Empty Quarter." Indeed, the western states from New Mexico to Montana offer seemingly unending open vistas, unspoiled by the crowding and dirtiness that too often characterizes the settlements of men. There is much more, however. The western states constitute a vast storehouse of energy. There is petroleum, especially new discoveries from the Overthrust Belt, low sulfur coal both for direct combustion and synthetic fuels, oil shale, and uranium. there is potential as well for geothermal and renewable resources; wind, biomass, and other forms of solar energy.

Eventual development of these resources is assured, but the when and how is not. Further, a number of science-based issues are currently of major concern. Four prominent questions deserve answers.

• Given the resource base in the west and elsewhere, what is the likely pace of western energy development, what technology is available for exploitation and conversion, and what are the costs?

• What is the likely long-term future for the energy-base region? Will exploitation follow practices least disruptive of the regional scene, or will major industrialization and vast sociological changes follow?

• What risks will energy development involve? Will it commit the allocation of other resources - water, land, ecosystems - in ways initially not anticipated?

•What are likely human consequences, and how are they best accommodated?

The portion of the symposium that this paper introduces addresses western energy resources and the more direct impacts of their exploitation. We seek to answer both the very general questions above and also questions as direct as what are the resources in the nation's treasurehouse? The primary source of answers to such questions will be found in three papers following this introduction: coal (by Gary B. Glass), oil shale (by James H. Gary and Arthur E. Lewis), and oil (by Richard B. Powers). As a corollary to these papers addressing the most important energy resources, it appears worthwhile to mention briefly energy resources that these authors do not address.

Uranium and Nuclear Energy

Despite Diablo Canyon and Three-Mile Island, nuclear energy can be important to our future. Recent estimates suggest that the states of New Mexico and Wyoming alone hold over 80 percent of the United States' uranium reserves; adding Arizona, Colorado and Utah raises this to about 90. Support for nuclear energy by the current administration should enhance activity, and indeed, the general attitude of the uranium mining industry is one of optimism over the long term. Growth in electrical demand, however, is slow, and the utility industry is faced with continuing uncertainties over nuclear cost and performance.

Data from a recent study of the global closed nuclear fuel cycle confirm that resources are available for nuclear growth, despite current low levels of activity and uncertainty in the industy (2). "Reasonably assured" supplies of uranium at a price to $130/kg uranium content are about 0.7 million metric tons, from the United States. More speculative reserves may add about 1.15 million metric tons. This represents a large fraction of the world-wide resource potential. A reactor fueled with uranium, when uranium is also recycled, requires about 4000 metric tons U_3O_8/GWe capacity over a 35 year life. Both surface and underground mining are widely practiced, with solution mining (in situ or heap mining) being developed as options.

Supply technology - mining, milling, conversion, enrichment, fuel fabrication, and construction of reactors - is well established with demand for services the critical bottleneck. What is more uncertain is the "back end" of the nuclear cycle: reprocessing of spent fuel, and waste management. Whether to reprocess or not and how hinge on decisions that extend far beyond the straightforward production of electricity from a light water reactor. There are concerns for plutonium produced as a byproduct, its security and ultimate use. There are concerns for high level radioactive waste; small in quantity but in part incorporating some risk extending longer than the lifetime of civilization's institutions. Various great deserts of the world are among natural sites with most favorable characteristics for disposal of highly radioactive waste (3). The Great Basin includes particularly well-suited opportunities (3, 4). Such areas are currently being investigated. There are also very large quantities of nuclear wastes with low radioactivity levels, but requiring safe disposal because of other noxious characteristics. Technically, these are likely in the end to prove the most difficult of all to manage.

These unanswered questions plus occasional ineptness shown by industry's technical management - demonstrated through malperformance - have established the current low rate of growth in use of nuclear energy. Eventually the resource is certain to be used, with impact on western areas. Now, however, a medium growth scenario predicts only about 550 GWe installed capacity in the western world through 2010 (2). Installed capacity was 122.5 GWe in 1981.

Geothermal Energy Resources

Geothermal energy, the natural heat of the earth, has been experiencing rapid development in areas such as The Geysers in northen California. There the steam-dominated hydrothermal resource has long been known, and electric power production soon may reach 1500 MWe.

World-wide, the amount of the geothermal resource to a depth of about 10 km is huge, representing about 2000 times the coal resources of the world (5). Economic concentrations for the generation of electricity, however, require at least about 150°C, available at depths less than about 3 km. Direct use (space or industrial heating) could prove of substantial importance. It is limited, however, by distribution of the resource away from population centers, and the fact that it cannot be transported until converted to electricity.

Development efforts concentrate along the western edge of the Basin and Range Province. These include Coso and Long Valley adjacent to the Sierra Nevada and Heber, East Mesa, and other locations in the Imperial Valley. The resource also is thought to extend across large areas within the Basin and Range Province, in central New Mexico and southern Colorado. Facilities are under development at scattered sites such as the Jemez Mountains of northern New Mexico and at Roosevelt Hot Springs in southwestern Utah. Community heating systems provide for direct use at several locations, including Boise, Idaho, and Klamath Falls, Oregon (5). Technically feasible proposals have been developed to use warm water effluent from electricity production as a heat source for several schemes: an integrated geothermal/algae biofuel facility, a proteinaceous algacultural operation, and an ethanol facility are examples (6).

Finally, of course, there is Yellowstone National Park. While development in the Park itself is prohibited, areas surrounding the Park offer excellent potential for low-temperature geothermal development. Leasing plans have been developed for the Island Park area of Idaho, adjacent to the Park. The government is proceeding with caution lest geothermal development outside of Yellowstone National Park adversely affect the thermal features within the Park (7).

Hydrothermal convection energy systems are believed capable of recovering perhaps 25 percent of the accessible energy originally present, from reservoirs meeting the criteria noted earlier. They thus can be a valuable additional resource even if limited to specific sites. We see that the present stage is one of relatively rapid exploration and steps toward development. There are, however, both generic and project-specific technical, environmental and institutional constraints that must be resolved before geothermal operations can be conducted (6). Many otherwise useful sources are likely to be plagued by corrosion and/or plugging of reinjection systems when dealing with more concentrated brines. An interesting sidelight is that local government for Inyo County in eastern California hopes to derive enough revenues from emerging alternatives such as geothermal energy to reduce taxation which otherwise would be needed (8).

Energy from the Wind

Energy from the wind is viewed as available at our fingertips, essentially free, non-polluting, and not subject to OPEC control. Of all the alternative sources which are renewable, wind appears most likely to experience rapid growth to large-scale commercial usage. The first successful wind park demonstrations could move it into full-scale development.

As with the geothermal resource, wind energy recovery at a commercial scale is limited to sites at which a substantial excess is present; wind must be consistent and high. In our region of interest such sites exist, but are likely to prove scattered - over the tops of ridges, and through mountain passes where air movement between basins is focused. Because economies of scale are possible with manufacture and utilization of thousands of devices, the economic potential still may prove large eventually. Like geothermal energy, wind resources are often found far from urban centers and not conveniently close to high voltage transmission lines. Thus though the potential is huge - perhaps it could stretch to about two percent of solar input (9) - the resource is seen throughout much of the west as of less immediate benefit than others.

California has been a leader in wind energy commercialization (10), adding state tax incentives to federal subsidies to spur activity. Projects underway in Solano County east of San Francisco, the Tehachapi Mountains, San Gorgonio Pass, and Boulevard, east of San Diego illustrate what may be in store for the larger western region this symposium addresses.

Despite some ideal characteristics, there are concerns showing potential conflict that may hinder wind energy development (11, 12). Chief among these are concerns for visual impact - visual clutter, to be caused by a virtual sea of whirling devices spread across a formerly open desert or mountain vista. Rotor safety is another issue; portions of blades have been thrown hundreds of meters, and there have been related deaths. A further concern is protection of wind rights - access to the wind for those wishing to use the resource on their property. Thus this "benign" source shows some need for regulation (and the concomitant slowing of entrepreneurial progress) as well as technological improvement.

Biomass

Since the land area of the western states is so very large and solar input generous, one can imagine much of the region converted to a giant harvesting operation for biomass energy products. The products could be burned directly for industrial heat or electricity. A perhaps more likely alternative would be production of a synthesis gas mixture and then methanol or petrochemical intermediates, eventually providing for both specialty products and transportation fuels.

The arid character of much of western lands substantially influences the kinds of biomass production likely to prove feasible. There are numerous plants uniquely adapted to this environment and the subject of current study. Several species of Euphorbia are under cultivation on a pilot basis (13). The buffalo gourd (Cucurbita foetidissima) is a perennial indigenous to Arizona and New Mexico (14). Guayule, a plant native to the Chihuahuan desert including Mexico and parts of Texas, can possibly be adapted within our region of interest (15). It produces latex, byproduct resins, and other carbon-containing materials, much like Euphorbia. Even the aspens characteristically treated as "weed-trees" in much of the mountainous west can be utilized as a source of energy (16).

If only less-productive land and natural rainfall are used, adverse environmental impact from such development may be quite limited. The resource as produced is scattered, however, and energy input, which may be very significant, is required to collect the raw material and transport it to conversion facilities (17). This is particularly true if waste or byproducts are to be utilized. Efficiency is gained through application of fertilizer and water, and use of land with better soil characteristics. At this point competition with food production would become direct and significant (18). The latter biomass approach does not appear to be a near-term useful alternative.

Technological Solar

There is a substantial basis for development of solar technology, of course. If the West industrializes, solar thermal energy could prove valuable for process heat, perhaps combined with production of electricity. The Barstow 10 MWe Solar 1 "power tower" soon may give some operating experience on which to substantially improve our estimates for the potential of this approach.

One option that has been studied is the hybrid solar thermal/coal power plant. This is an option that could reduce some adverse impacts of large-scale use of coal on air quality in western regions where clean air is highly prized. Although particulates from coal and nitrogen oxides are reduced through this combination, other problems due to the character of the interactive system may ensue (19). When very high temperatures are attained at a solar receiver face in order to meet efficiency goals, NOx is produced just as with fossil plants, though at a much less intense pace (20). Thus highly technical development can make use of the western solar resource, but it entails system design, use of materials, and system operation that are relatively sophisticated and costly. There is potential for off-normal behavior - complex systems running out of control - incorporating different hazards than those we have become accustomed to and managed successfully in the past (21).

Solar troughs, from one viewpoint a scaled-down approach to solar thermal use, may have more immediate application than the higher technology devices. Southern California Edison (a private utility) plans to construct and operate a 15 MWe solar trough electric generating facility in the Mojave desert adjacent to Solar 1 (22). Because conditions need not be as extreme as for solar tower design, some aspects of complexity and concern are diminished in the solar trough approach. For example, lower peak temperatures may reduce materials demands. At the same time, working fluid and thermodynamic cycle requirements at lower temperatures may make efficiency more difficult to achieve. The land committed to this use remains a substantial factor. There is the potential for influencing vast surrounding areas if activity is intensive enough to justify the capital investment needed.

For many years enthusiasts have tabbed solar cells for direct conversion of sunlight to electricity as the likely best solar technology in the long run. A prime advantage is that small-scale and individual systems can be effective, achieving an "appropriate technology." The continuing problem has been to obtain sufficiently high conversion efficiency within the solar cell to produce a low cost material (and conversion system) through mass production and economies of scale. "Breakthroughs" of one kind or another have been announced with some regularity, only to sink after realistic assessment to the level of a healthy but still limited improvement in the state-of-the art. Nevertheless, gradual progress is taking us forward. The long-ago established goal of ten percent operating efficiency has been achieved for an amorphous silicon solar cell (23). This is termed "comparable to running the first four-minute mile" by a spokesman for the developer. Unfortunately, equivalent improvements in some equally critical portions of an energy delivery system are not as likely: installation, maintainability, power conditioning, and interconnection with backup supplies or to feed into the grid.

Solar ponds represent yet another means by which to harvest energy from the sun. The concept is not new (24). Sunlight heats one fluid, and then most likely after contact through an intermediate heat transfer system, heat from a second fluid drives electric power generation. Plans are well advanced to try out this scheme in a portion of the Salton Sea in the southeastern desert area of California (25). Research currently in progress at the University of California, Los Angeles, addresses several pertinent environmental concerns. Salt concentration profiles are mandated by the technology and they must be held very close to design levels. The salts used can affect waterfowl, fisheries, and other wildlife. Studies address possible impacts of changes in Salton Sea salinity, concentration of specific ions, and other factors. A complication is that agricultural return flows now supply most of the water entering this lake. These contain not only runoff nutrients, but pesticide residues and other elements that may be of concern in the changed environment. Introducing solar pond use brings in another dimension, adding greatly to complexity and the task of responsible resource management.

Utah, a state central to the energy resource domain that is the focus of this symposium, should be alert to the outcome of these experiments. The Great Salt Lake might pose an extremely attractive solar energy opportunity.

References

1. Garreau, J. The Nine Nations of North America. (Houghton-Mifflin, Boston, 1981). 427 p.
2. Environmental Science and Engineering. Logistics of the Nuclear Fuel Cycle. Report prepared for the Electric Power Research Institute, UCLA (in press).
3. Environmental Science and Engineering. Evaluation of Great Deserts of the World for Perpetual International Radiowaste Storage. EPRI Report ESC-2277-LD, Contract TPS80-753 (1982).
4. Winograd, I.J. "Radioactive Waste Storage in the Arid Zone". EOS (AGU Transactions) 55(10):884-894 (1974). [Discussion in EOS 57(4):178, 215-216.]
5. Muffler, L. J. Patrick. "Geothermal Energy and Geothermal Resources of Hydrothermal Convection Systems". Rubey Colloquium Lecture, UCLA. Rubey Colloquium Volume III, Prentice-Hall (in press).
6. Thomas, Terry Robert. "Environmental Planning for Geothermal Resource Exploration, Development and Utilization". UCLA, dissertation report for Doctor of Environmental Science and Engineering (1982).
7. U.S. Department of Agriculture. Final Environmental Impact Statement of the Island Park Geothermal Area, Idaho, Montana, Wyoming. U.S. Forest Service, Ogden, Utah (1980). 280 p.
8. Anon. "DWP Caught in the Middle on Air Pollution Permits". Inyo Register, Bishop, California, section A, p. 2. (June 17, 1982).
9. Gustavson, M.R. "Limits to Wind Power Utilization". Science 204(4388): 13-17 (1979).
10. Ginosar, Matania. "A Proposed Large-Scale Wind Energy Program for California". Energy Sources 5(2): 141-169 (1980).

11. Riverside County/U.S. Bureau of Land Management. Draft Environmental Impact Report/Statement. DEIR/EIS #158, by Wagstaff and Brady Consulting Engineers (March 1982).
12. Riverside County/U.S. Bureau of Land Management. San Gorgonio Wind Resource Study. Final Environmental Impact Report/Statement. EIR/EIS #158 (SCH 81101903, EPA DEIS No. 82-8). (July 1982).
13. Calvin, Melvin. "Petroleum Plantations for Fuel and Materials". BioScience 29(9): 533-538 (1979).
14. Morgan, Robert P., and Eugene B. Schultz, Jr. "Fuels and Chemicals from Novel Seed Oils". Chem. and Engr. News 59(36): 69-77 (1981).
15. Campos-Lopez, Enrique, et al. "The Rubber Shrub". Chemtech 9(1): 50-57 (1979).
16. Myerly, R.C., et al. "The Forest Refinery". Chemtech 11(3): 186-192 (1981).
17. Pimentel, David, et al. "Biomass Energy from Crop and Forest Residues". Science 212(4499): 1110-1115 (1981).
18. Dritschilo, W., et al. "Energy vs. Food Resource Ratios for Alternative Energy Technologies". Energy 8(4): 255-265 (1983).
19. Perrine, R.L. "Some Environmental Considerations in Siting a Solar/Coal Hybrid Power Plant", in Beyond the Energy Crisis: Opportunity and Challenge, Vol. II, 623-630 (Pergamon Press, 1981).
20. Perrine, R.L., et al. The Potential Prediction of Air Pollutants Near STPS Surfaces. Report UCLA 12-1313. Laboratory of Biomedical and Environmental Sciences, UCLA (1981).
21. Perrine, Richard L. "Environmental Concerns for Off-Normal Events with Solar Thermal Power Systems". Second U.S. DOE Environmental Control Symposium, Reston, Virginia, March 17-19 (1980).
22. Anon. "Edison to Buy Power from Solar Trough". Inyo Register, Bishop, California, section B, p. 1 (July 29, 1982).
23. Anon. "Solar Cells Reach 10% Efficiency". Chem. and Engr. News 60(31): 18 (1982).
24. Tabor, H. "Solar Ponds". Solar Energy 7: 189 (1963).
25. Sciarrotta, Terry. Research Division, Southern California Edison Co., personal communication (October 1981).

6. Wyoming, an Example of Western Coal Development

Occurrence of Coal

Based on the most current estimate by the U.S. Geological Survey, northern and southern Rocky Mountain States contain an estimated 2.7 trillion tons of coal or about 67 percent of the nation's four trillion tons of coal resources (1). Of the nation's reserve base, which is the potentially recoverable portion of the coal resource, the Rocky Mountain area has 112.6 billion tons of coal that are potentially minable by underground techniques (38 percent of U.S. total) and 86.4 billion tons that are potentially minable by surface mining methods (63 percent of U.S. total). Eighty percent of this surface minable western reserve base, incidentally, occurs in Montana and Wyoming (1).

Wyoming, alone, accounts for 35 percent of the coal resources and 26 percent of the reserve base in the Rocky Mountain area (1). Because of this and Wyoming's geologic and physiographic settings, a discussion of the coal deposits and coal mining in Wyoming provides an excellent overview of western coals and their current development.

Wyoming's coal fields fall into the same two coal-bearing provinces as coal deposits of the other Rocky Mountain states. The coals in northeastern Wyoming are within the Northern Great Plains Province while all other coal deposits of the State are in the Rocky Mountain Province. Wyoming's coal-bearing areas are further divided into 10 major regions, basins, or fields, which underlie more than 40,000 square miles or approximately 41 percent of the State and which collectively contain almost 24 percent of the nation's coal resources under less than 6,000 feet of overburden (Fig. 1).

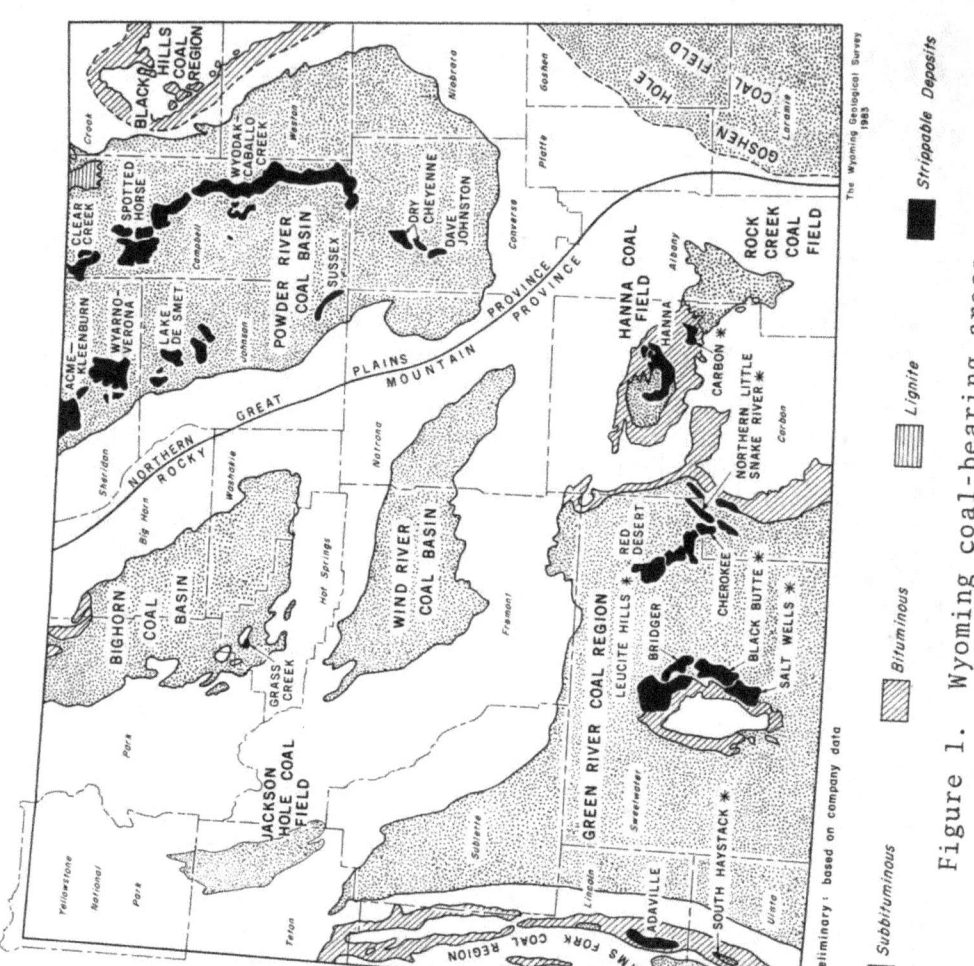

Figure 1. Wyoming coal-bearing areas.

Major coal-bearing areas are further subdivided into 45 individual coal fields (Fig. 2). These coal fields were originally defined by numerous published reports, which quite frequently did not give precise boundaries for a given field. For this reason, the field boundaries on Figure 2 have been revised so that (1) field boundaries do not overlap and (2) major coal-bearing areas do not have regions which aren't included in a defined field. It is noteworthy that, at least locally, coals also occur outside the defined coal-bearing areas and coal fields.

Geologic Coal-Bearing Formations

As in most other Rocky Mountain states, Wyoming's coals occur in rock sequences deposited during either the Cretaceous Period (some 66-135 million years ago) or during the younger Tertiary Period (38-66 million years ago). During both these periods, depositional environments and climates were at least periodically well suited to the development of densely vegetated swamps. Peats that accumulated in these swamps have since been transformed into the nearly trillion tons of coal that still underlie 41 percent of the State.

Geologic formations that contain these coals characteristically are thick; generally 700-7,000 feet in thickness. While the Cretaceous formations normally exhibit gradual, regional thickening or thinning across the State, the thicknesses of the various Tertiary formations vary from basin to basin. These variations are more a result of local tectonic and depositional events that affected each of the coal-bearing areas than they are related to the larger regional events that marked the Cretaceous Period.

The most widespread coal-bearing rocks in Wyoming are Cretaceous in age and these rocks usually crop out only as narrow bands of upturned strata along the margins of the larger structural basins and uplifted areas of the State. They also crop out as irregularly exposed, linear bands in the thrust belt of western Wyoming (Fig. 1). Relatively flat-lying Tertiary rocks, on the other hand, occupy the central portions of most of the coal-bearing areas where they overlie the older Cretaceous rocks. Even the Tertiary rocks often exhibit steeper dips as they approach the margins of the coal-bearing basins and regions.

Cretaceous and Tertiary coal-bearing formations contain numerous coals that are separated from one another by as little as a few inches of shale or claystone to hundreds of

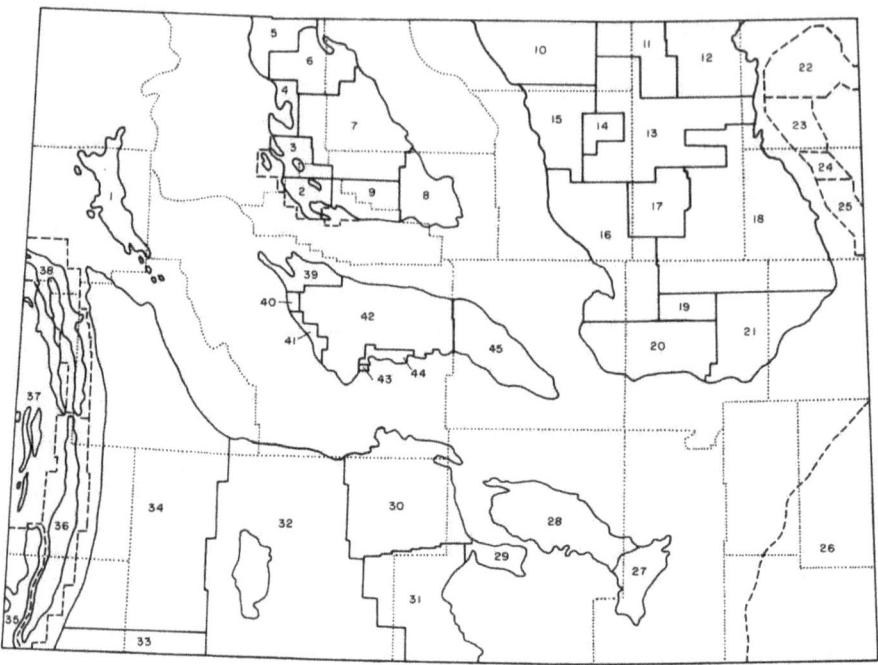

1. Jackson Hole	24. Skull Creek
2. Grass Creek	25. Cambria
3. Meeteetse	26. Goshen Hole
4. Oregon Basin	27. Rock Creek
5. Silvertip	28. Hanna
6. Garland	29. Kindt Basin
7. Basin	30. Great Divide Basin
8. Southeastern	31. Little Snake River
9. Gebo	32. Rock Springs
10. Sheridan	33. Henry's Fork
11. Spotted Horse	34. La Barge Ridge
12. Little Powder River	35. Evanston
13. Powder River	36. Kemmerer
14. Barber	37. Greys River
15. Buffalo	38. McDougal
16. Sussex	39. Muddy Creek
17. Pumpkin Buttes	40. Pilot Butte
18. Gillette	41. Hudson
19. Dry Cheyenne	42. Beaver Creek
20. Glenrock	43. Big Sand Draw
21. Lost Spring	44. Alkali Butte
22. Aladdin	45. Arminto
23. Sundance	

Figure 2.   Coal fields of Wyoming.

feet of rock that may vary from coarse sandstone or conglomerate to siltstones, claystones, and shales. Although the Cretaceous coals interspersed in these rocks are generally less than ten feet in thickness, a few Cretaceous coals are 30 to 100 feet thick in westernmost Wyoming (2).

The Tertiary coals, which were deposited during the Eocene and Paleocene epochs, often exceed ten feet in thickness with 30 to 80 feet thick coals common. Locally, at least one Tertiary coal is 220 feet in thickness (3, 4).

Figure 3 shows Wyoming's coal-bearing formations arranged by coal-bearing area. The oldest coal-bearing formation in Wyoming is the Lower Cretaceous Lakota Conglomerate. This formation contains at least one minable coal in the Black Hills Coal Region.

The Bear River Formation of Lower Cretaceous age is the next younger coal-bearing rock unit above the Lakota. The coals in the Bear River Formation are very local in extent and have only been reported in the Hams Fork Coal Region.

Separated from the Bear River Formation by a marine shale, the overlying Upper Cretaceous Frontier Formation (variously mapped as the Blind Bull Formation) contains numerous fairly thick, persistent coals in western Wyoming. Elsewhere, the Frontier coals apparently are thin, shaly, and of very limited extent.

The oldest widespread coal deposits in Wyoming are found in the Upper Cretaceous Mesaverde Group or its western equivalent, the Adaville Formation. These rocks contain numerous thick to moderately thick coals in the Hams Fork and Green River regions. Mesaverde coals are less numerous and apparently thinner and more local in extent throughout the rest of the State. The Mesaverde disappears before it reaches half way across the Powder River Basin, and only contains very thin coals on the southern flank of that basin.

The Meeteetse Formation was deposited at about the same time as the Lewis Shale. This formation is recognized in the Wind River and Bighorn basins where it contains some coals, most of which are thin and discontinuous (5, 6).

Lance Formation coals are the youngest Upper Cretaceous coals in Wyoming. Although the Lance and equivalent aged rocks contain coal throughout the State, Lance coals are best developed in southern Wyoming. There they are numerous, but seldom more than 10 feet in thickness.

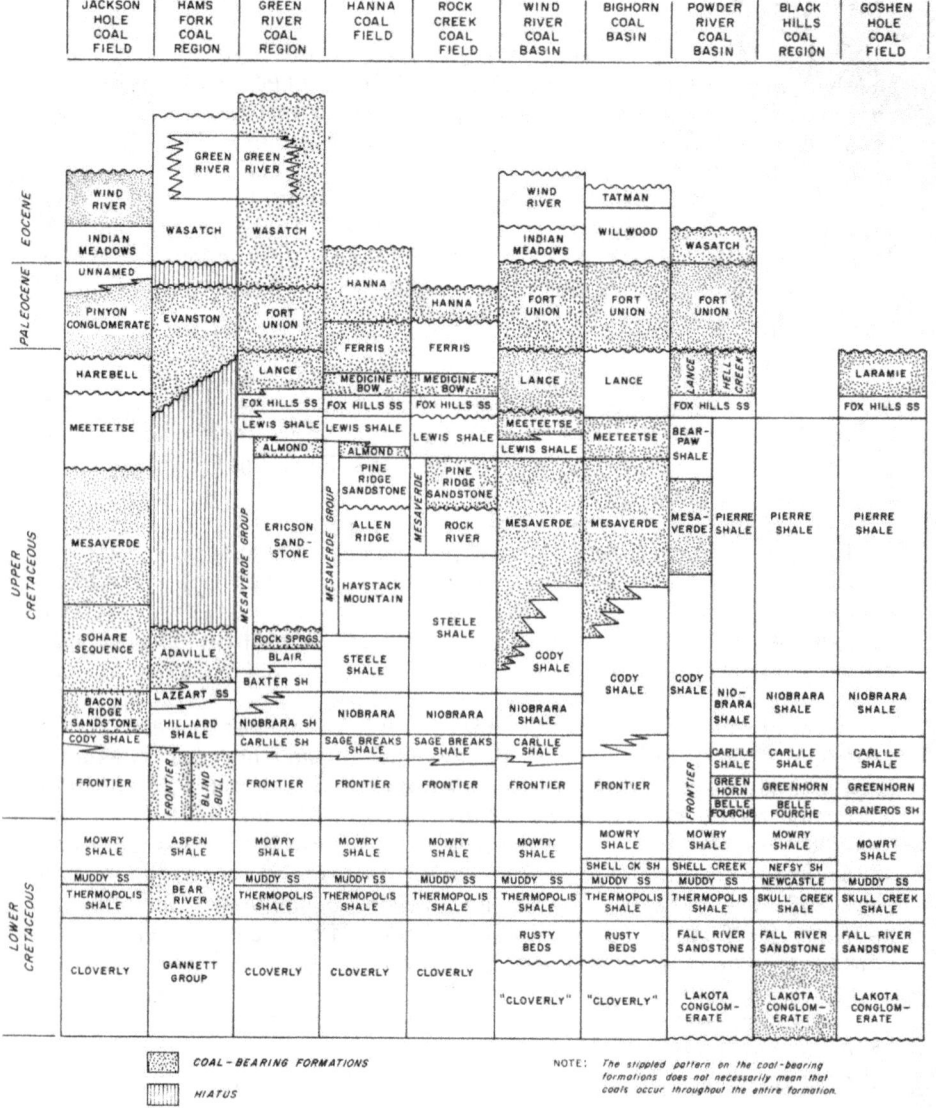

Figure 3.   Major coal-bearing formations of Wyoming.

Paleocene rocks (variously mapped as the Fort Union, Pinyon Conglomerate, or Evanston formations) crop out in all but the Black Hills Region and Goshen Hole Field. Paleocene rocks invariably contain coals although they are most prolific in the Powder River Basin and Hanna Field (3, 4, 7). While there are more Paleocene coals in the Ferris and Hanna formations of the Hanna Field, the Fort Union coals of the Powder River Basin are often two to three times as thick as the thickest coals in the Hanna Field, which are as thick as 36 feet.

The Eocene Wasatch Formation is the youngest coal-bearing rock unit of economic importance in Wyoming. At least in the Powder River Basin, the Wasatch rivals the older Paleocene rocks in both the number of coals it contains and in their thickness (3, 4). In fact, the Wasatch contains the thickest coal in Wyoming, the 220 feet thick Lake De Smet coal. Wasatch or equivalent aged coals are also abundant and moderately thick in the Great Divide Basin of the Green River Region and in the Hanna Field. Elsewhere in the State, they are thinner and less persistent than many older coals.

In general, Wyoming coal measures like those in other Rocky Mountain states, are situated in broad, asymmetrical, synclinal basins between various ranges of the Rocky Mountains. Except for those coal beds that are tilted against the Rock Springs Uplift in the central portion of the Green River Region, most of the State's coals are relatively flat-lying in the more central portions of the basins. Steeper dips and significant folding are common at some basin margins as well as on the flanks of mountain ranges.

While the Hams Fork Region and Hanna Field exhibit the greatest structural complexity, the Powder River Basin shows the least. Faulting is most common in the southern and western coal regions, but it is not restricted to those areas.

Wyoming Coal Quality

The rank of Wyoming coal is lignite to high volatile A bituminous. Lignites, however, are restricted to a small area in the northeastern corner of the Powder River Basin. That occurrence represents a southern extension of the Tertiary lignite deposits of Montana and North Dakota.

Subbituminous coals which are all either Tertiary or Late Cretaceous in age, are found in all the coal-bearing areas of the State except the Black Hills Region. Usually, the subbituminous coals occupy the more central parts of the

TABLE 1. Characteristics of Wyoming coals that are currently mined.

Coal-bearing Area	Apparent Rank	Bed Thickness (feet)	Moisture (%) (As-received basis)	Ash (%) (As-received basis)	Sulfur (%) (As-received basis)	Heat Value (Btu/pound) (As-received basis)	Hardgrove Grindability Index
Powder River Coal Basin							
Northwestern portion	Subbituminous	Range: 10-57 Average: 40	Range: 14.5-26.0 Average: 21.5	Range: 3.1-8.2 Average: 4.4	Range: 0.3-1.0 Average: 0.4	Range: 9,000-10,410 Average: 9,600	Range: 39-50 Average: 41
Eastern portion	Subbituminous	Range: 50-100 Average: 75	Range: 21.1-36.9 Average: 29.8	Range: 3.9-12.2 Average: 6.0	Range: 0.2-1.2 Average: 0.5	Range: 7,420-9,600 Average: 8,220	Range: 49-55 Average: 53
Southern portion	Subbituminous	Range: 17-38 Average: 35	Range: 19.5-29.3 Average: 22.2	Range: 6.6-15.7 Average: 11.4	Range: 0.4-0.7 Average: 0.6	Range: 7,610-8,870 Average: 8,180	Range: 30-35 Average: 35
Hanna Coal Field	Subbituminous and bituminous	Range: 5-38 Average: 12	Range: 6.0-20.8 Average: 11.9	Range: 3.8-21.3 Average: 8.7	Range: 0.2-2.1 Average: 0.8	Range: 8,310-12,600 Average: 10,310	Range: 43-110 Average: 55
Green River Coal Region	Subbituminous	Range: 5-30 Average: 15	Range: 17.0-20.5 Average: 20.5	Range: 5.9-10.0 Average: 9.7	Range: 0.4-0.8 Average: 0.5	Range: 9,270-10,000 Average: 9,350	Range: 79-82 Average: 80
Hams Fork Coal Region	Subbituminous	Range: 5-90 Average: 16.5	Range: 15.4-28.6 Average: 20.8	Range: 1.5-8.9 Average: 3.4	Range: 0.2-1.8 Average: 0.6	Range: 7,920-10,550 Average: 9,860	Range: 41-87 Average: 51
Bighorn Coal Basin	Bituminous	Range: 8-38 Average: 20	Range: 10.7-12.8 Average: 12.3	Range: 5.0-9.4 Average: 7.4	Range: 0.3-0.6 AverageL 0.4	Range: 10,730-11,246 Average: 10,970	Range: - Average: -

coal-bearing areas although, in the case of the Hams Fork Region, subsequent erosion has relegated even the younger subbituminous coal-bearing rocks to narrow bands between faults and eroded folds. In several basins the subbituminous coals are buried beneath great thicknesses of rock that do not contain coals.

Most of Wyoming's bituminous coals crop out as narrow bands in the Hams Fork Region, around the Rock Springs Uplift, along the eastern edge of the Green River Region, and on the periphery and eastern half of the Hanna Coal Field. The Mesaverde coals in the north end of the Bighorn Basin and the Lower Cretaceous coal of the Black Hills Region are also bituminous. Bituminous coals are actually much more widespread than these outcrops would suggest because a great portion of them lie deeply buried (some are well over 10,000 feet deep).

While the older coal beds in any given field are generally higher in rank than the younger beds, the rank of individual beds in a field also seems to increase toward the troughs of the structural basins. Both of these variations in rank can be attributed to increases in depth of burial.

In the Hanna Coal Field, however, the rank of Tertiary age coals increases eastward across the coal field (7). As a result, Paleocene and Eocene coals of the Hanna Formation exhibit higher apparent ranks than the Lower Cretaceous and Paleocene coals that crop out in the western half of the Hanna Field. This variation in rank is attributed more to variations in heat flow within the field than simply to greater depth of burial.

Moisture, volatile matter, and fixed carbon contents of coals vary widely across the State in response to variations in rank (Table 1). The bituminous and higher ranked subbituminous Cretaceous and Tertiary coals are very similar on an as-received basis. Moisture contents are less than 15 percent, volatile matter contents are between 30-40 percent, and fixed carbon contents are greater than 40 percent. In contrast, the lower ranked subbituminous Tertiary coals of the Green River Region and the Powder River Basin have as-received moisture contents between 20-30 percent and about equal volatile matter and fixed carbon contents.

Ash and sulfur contents of mined Wyoming coals are characteristically low and are related to the depositional histories of the coals rather than to any differences in rank (Table 1). Consequently, variation in ash contents, in particular, are quite irregular. As-received ash contents

are from a few percent to more than 50 percent for some
unmined coals in response to the volume of inorganic debris
entering the original peat swamp.   Published analyses of
various coals sampled across the State, however, suggest that
a typical, persistent Wyoming coal of minable thickness
contains less than 10 percent ash.

The sulfur content of Wyoming coals is quite variable.
For example, relatively high sulfur coals occur in the
Wasatch Formation in the Red Desert area of the Green River
Coal Region and in the uppermost Hanna Formation coals of the
Carbon Mining District of the Hanna Coal Field.   Although
none of these coals is mined, exploration has shown as-
received sulfur contents as high as 7.2 percent in the Red
Desert area (8) and 8.7 percent in the Carbon Mining District
(9).

Published   analyses   of   Wyoming's   coals   generally
highlight   the   lower   sulfur   coals.   For   instance,   most
published analyses of the State's Cretaceous coals show 0.9 -
2.0 percent sulfur on an as-received basis, compared to 0.3 -
0.9 percent sulfur in the State's Tertiary coals.   In some
cases, relatively low sulfur contents are the result of an
old practice of removing impurities before analysis.   In
other cases, analyzed samples are coming from coal mines
where contract specifications require the exclusive mining of
low sulfur coals.   For this reason, mined Wyoming coals now
average 0.5 percent sulfur and rarely exceed 1.0 percent
sulfur on an as-received basis.

Heat values of currently mined coals, like moisture
contents, vary widely across Wyoming (Table 1).   In the
southern half of the State, as-received heat values average
10,500 Btu/lb.   They typically are between 9,270 and 11,700
Btu/lb.   Some younger coals, mined at the Jim Bridger strip
mine in southwestern Wyoming, account for the lower heat
values.   These coals average only 9,350 Btu/lb.   While heat
values of coals mined in western Wyoming average 9,600
Btu/lb., the heat values in northeastern Wyoming are much
lower.   Although heat values in northeastern Wyoming coals
are from 9,300 Btu/lb. in the Sheridan area to 7,550 Btu/lb.
in Converse County, they average only 8,300 Btu/lb.

In Wyoming coals, major elements, which make up more
than 0.1 percent of a coal, are silicon, calcium, aluminum,
iron, and magnesium, usually in that order of abundance.
Common minor elements, which account for less than 0.1
percent of a coal, are potassium, sodium, titanium,
phosphorous, chlorine, and manganese, in that order.

Silicon, aluminum, calcium, and iron all can be present in concentrations greater than 2 percent; silicon concentrations have exceeded 5 percent in some samples (Table 2). The minor elements rarely exceed 0.1 percent and are best described in parts per million (10).

Table 2. Major and Minor Elements in 48 Wyoming Coal Samples in Percent on a Whole-Coal Basis (11, 12).

	Wyoming Range	Wyoming Average	U.S. Average
Silicon	0.41-5.50	1.70	2.6
Calcium	0.13-2.10	0.75	0.54
Aluminum	0.17-2.50	0.72	1.4
Iron	0.15-2.100	0.51	1.6
Magnesium	0.026-0.340	0.17	0.12
Potassium	0.005-0.370	0.063	0.18
Sodium	0.003-0.190	0.044	0.06
Titanium	0.001L-0.130	0.038	0.08
Phosphorous	0.0021L-0.044	0.0121	--
Chlorine	0.004-0.026L	0.010L	--
Manganese	0.0007L-0.0492	0.004	0.01

Based on some recent analyses, Wyoming's Cretaceous and Tertiary coals show similar concentrations of all the major and minor elements except calcium and sodium (11). The concentrations of calcium and sodium in the Tertiary coals are usually four to five times the concentrations normally reported in the Cretaceous coals of Wyoming.

Of 30 trace elements generally recognized in analyses of Wyoming coals, concentrations of the various elements vary from a high of 1,000 parts per million to a low of 0.004 parts per million (10). If these elements are grouped according to their average concentrations on a whole-coal basis, 6 elements average less than 1 part per million, 12 elements are between 1-5 parts per million, 11 elements are between 5-100 parts per million, and one element, barium, averages more than 300 parts per million (Table 3).

Table 3.   Average Trace Element Concentrations in 48 Wyoming Coal Samples in Parts Per Million on a Whole-Coal Basis (11,12)

Element	Wyoming Average	U.S. Average
Barium	300.	150.
Strontium	100.	100.
Boron	70.	50.
Fluorine	70.	74.
Cerium	< 20.	--
Zinc	17.9	39.
Vanadium	15.	20.
Zirconium	15.	30.
Neodymium	< 15.	3.
Copper	8.	19.
Chromium	7.	15.
Lanthanum	< 7.	--
Nickel	5.	15.
Yttrium	5.	10.
Lithium	4.6	20.
Gallium	3.	7.
Arsenic	< 3.	15.
Lead	< 3.	16.
Thorium	2.7	4.7
Cobalt	2.	7.
Germanium	< 2.	--
Niobium	1.5	--
Scandium	1.5	3.
Molybdenum	1.0	3.
Uranium	< 0.9	1.8
Selenium	< 0.8	4.1
Ytterbium	0.5	1.
Antimony	< 0.4	1.1
Cadmium	< 0.15	1.3
Mercury	0.10	0.18

With the exception of zinc, which averages 17.9 parts per million, the more common metals -- copper, cobalt, nickel, and lead -- all occur in concentrations from 1-5 parts per million. While the potentially dangerous elements, arsenic and molybdenum, also average 1-5 parts per million, selenium, cadmium, and mercury normally occur in concentrations of less than one part per million (10).

Based on published analyses, Tertiary coals in Wyoming appear to contain higher concentrations of 26 of these trace elements than do the Cretaceous coals. Only boron, beryllium, fluorine, and germanium are higher in the Cretaceous coals (11, 12).

Coal Production

Except for an eleven-year interval after World War I, Wyoming's coal production for the years 1910 through 1945 remained above 6 million tons annually. After 1945, production fell to a record low of 1.6 million tons by 1958. This decline followed World War II, and more importantly, the railroad's change from steam locomotives to diesel engines.

Renewed interest in Wyoming's coal resources began in the late 1960's as power plant demands for inexpensive, low sulfur coals increased. Because of this demand, Wyoming's coal companies have set new production records every year since 1972.

Wyoming's 1980 tonnage set yet another record at 94.48 million tons (Fig. 4). With this tonnage, Wyoming remains the largest coal-producing state in the Rocky Mountains and the third largest in the nation. In 1981, production will probably exceed 108 million tons. Present indications suggest annual tonnage will be about 157 million tons in 1985 and 176 million tons by 1990 (13). These projected increases are principally to satisfy the electric power market and include no coal gasification plants. At these rates of increase, Wyoming's coal production in the 15-year period between 1971 and 1985 will be more than twice the production from 110 years of mining prior to these years or approximately one billion cumulative tons.

Coal production in Wyoming was dominated by underground mining until 1954. In that year strip mining tonnage barely exceeded that of the underground mines. Since then, however, strip mining has become the dominant mining method and now accounts for more than 98 percent of Wyoming's annual production. Conversely, underground mining has slipped to 2

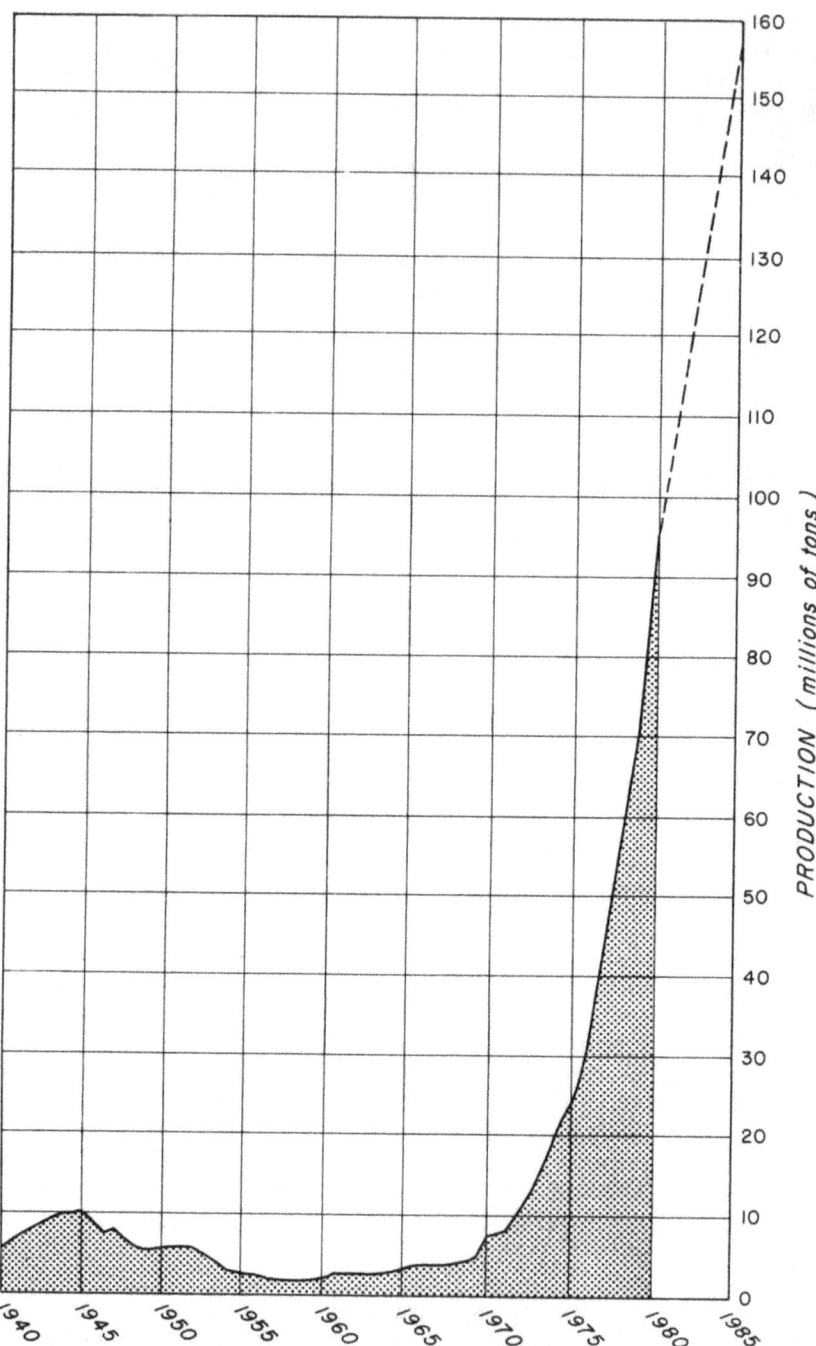

Figure 4. Wyoming coal production 1940-1980, with projections to 1985.

percent of the annual tonnages mined. An estimate of the cumulative total coal production from Wyoming, to January 1, 1981, is 823.3 million tons of which 47 percent or 389.1 million tons came from underground mines and 53 percent or 434.2 million tons came from surface mines.

In 1980, twenty-one coal mining companies produced 94.48 million tons of coal. These companies operated 23 strip mines and 3 underground mines. Two of the underground (deep) mines accounted for 86 percent of the underground tonnage of 1.63 million tons. The eight strip mines that produced in excess of 4 million tons, accounted for 70 percent of the strip mine tonnage. Amax Coal Company's Belle Ayr strip mine remained the largest coal mine in the United States in 1980, producing 16.1 million tons, which is 2 million tons less than its record production of 18.1 million tons in 1978.

In 1980, about 79 percent or 74.9 million tons of the coal mined in Wyoming went to power plants in at least 16 other states, located as far south as Texas and Louisiana, and as far east as Indiana, Illinois, and Ohio. Another 18 percent, or 16.5 million tons was burned in Wyoming power plants (14). Industrial customers used most of the remaining three percent. The beet sugar, cement, and phosphate industries were the major out-of-state industrial users, with about 1.25 million tons shipped to various plants throughout the Pacific Northwest, Rocky Mountain, and Mid-continent regions. Another 1.6 million tons were used to manufacture trona, cement, bentonite, and synthetic coke in Wyoming. In-state and out-of-state retail sales of coal were each about 100,000 tons.

Wyoming's remaining in-place coal resources between 0 and 3,000 feet of overburden are now estimated at 138,197,810,000 short tons (10). Approximately 11 percent of these resources are bituminous coal and 89 percent subbituminous coal. These resources, however, are based on only 47 percent of the known or probable coal-bearing land in Wyoming as they are limited to mapped and explored areas. When an estimate of the resources of the previously omitted 53 percent of the State's coal-bearing land is added to the mapped and explored estimate, the U.S. Geological Survey estimates that Wyoming's remaining resources under less than 3,000 feet of overburden increase to 838,197,810,000 tons. Wyoming's remaining resource figure becomes 938,197,810,000 tons when the overburden category is extended to 6,000 feet. In the 0 to 6,000 feet overburden category, Wyoming has the largest in-place coal resources in the nation at approximately one trillion tons (1, 10).

Wyoming's coal reserve base is approximately 56 billion tons (10). Of this, 26 billion tons is strippable. This strippable reserve base is limited to a few mapped and explored areas and is only a small portion of wyoming's potentially strippable coal, which could easily total ten times the known reserve base.

References and Notes

1. P. Averitt, U.S. Geol. Surv. Bull. 1412 (1975).
2. G.B. Glass, Wyo. Geol. Assoc. Guidebook 29th Ann. Field Conf. (1977), pp. 689-706.
3. _____, Wyo. Geol. Assoc. Guidebook 28th Field Conf. (1976), pp. 209-220.
4. _____, Ed., Geol. Surv. Wyo. Public. Info. Cir. 14 (1980).
5. _____, K. Westervelt, and C.G. Oviatt, Wyo. Geol. Assoc. Guidebook 27th Ann. Field Conf. (1975), pp. 221-228.
6. _____, and J.T. Roberts, Wyo. Geol. Assoc. Guidebook 30th Ann. Field Conf. (1978), pp. 363-377.
7. _____, Geol. Surv. Wyo. Rpt. Invest. 22 (1980).
8. H. Masursky, U.S. Geol. Surv. Bull. 1099-B (1962), pp. B1 - B52.
9. G.B. Glass, Geol. Surv. Wyo. Rpt. Invest. 16 (1978).
10. _____, Geol. Surv. Wyo. Public Info. Circ. 9 (1978).
11. _____, Geol. Surv. Wyo. Rpt. Invest. 11 (1975).
12. V.E. Swanson, et al., U.S. Geol. Surv. Open-File Rpt. 76-468 (1976).
13. G.B. Glass, Geol. Surv. Wyo. Public Info. Circ. 12 (1980).
14. M.B. McNair, U.S. Dept. Energy, Rpt. DOE/EIA-0125 (80/4Q) (1981).

Richard B. Powers

7. Significance of Recent Oil and Gas Discoveries in the Western Thrust Belt of the United States

Introduction

Because of a series of significant large oil and gas field discoveries in southwestern Wyoming and northcentral Utah during the past eight years, the name "Overthrust" has acquired an almost magical connotation. Some have gone so far as to suggest that the western thrust belt contains an underground river of oil extending from Alaska to Acapulco (1). If this were true, the question of our nation's future supplies of oil and gas would be perhaps less critical.

The western thrust belt is a narrow (35-125 mi wide) belt of complex thrust faults and folds that stretches from the Canadian border at the north, to Mexico at the south. It forms the middle part of the greater Cordilleran thrust complex which is generally considered to be a single, but geologically variable, tectonic element that extends for nearly 5,000 miles from Alaska to Central America. It is also part of the vast system of mountain ranges and basins that make up the North American Cordillera, whose topography varies from the rugged, snow-covered peaks of Alaska's Brooks Range to arid desert areas of Arizona and Mexico. Parts of this thrust fold feature have been explored and drilled for oil and gas resources over a period of some 90 years (Fig. 1). However, the United States' part has not been drilled to the same extent as other areas in the Rocky Mountains which have preserved adequate thicknesses of sedimentary rocks and geologic histories that are favorable for oil and gas accumulations.

Many geologists considered that the thrust belt region has had too long and complicated a history of multiple tectonic deformation, uplift and erosion to harbor

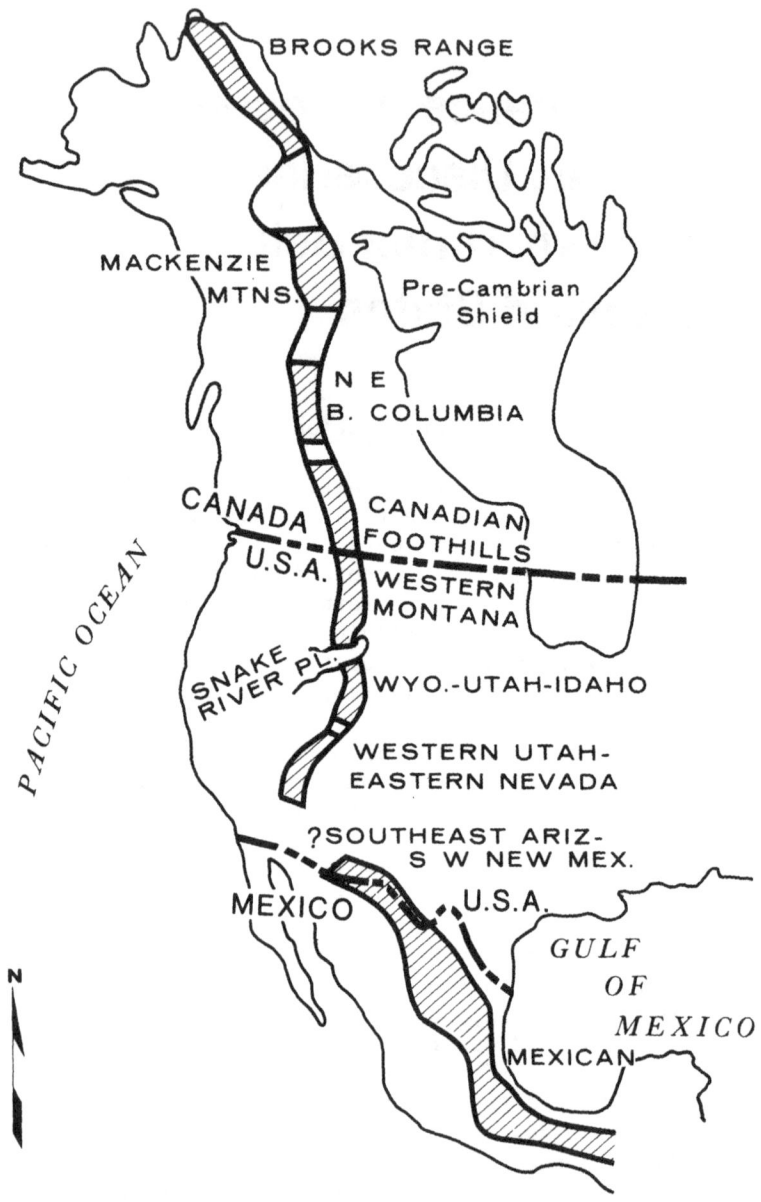

Figure 1. Sketch map of the greater Cordilleran thrust belt of western North America showing eight recognized major thrust salients (diagonal lines). Continuation of the thrust belt in southwest Arizona is questionable (?) owing to lack of data. The area of study is within the Wyoming-Utah-Idaho salient. Modified from Hayes, (2); and Powers (3).

significant oil and gas reserves, even though known rock thicknesses and facies seem favorable for oil and gas generation and migration into trapping structures (4). In spite of these favorable geologic characteristics, the thrust belt was once considered to be a "driller's graveyard", as attested to by the numerous dry holes that dotted this extensive region. In fact, more than 100 significant wildcat dry holes were drilled in the Wyoming-Utah-Idaho thrust belt alone before the first important oil and gas accumulation was discovered at the Pineview field in Summit County, Utah, in early 1975.

Early Exploration History

Exploration for oil and gas in frontier regions such as the thrust belt ordinarily involves mapping the structure, determining thickness and extent of rock outcrops on the surface, and analyzing rock samples for reservoir or source-rock quality, and then comparing these factors to analogous, productive provinces in the region. In the initial phase of exploration, however, perhaps the most positive evidence for a potential hydrocarbon province is the presence of active oil and gas seeps or asphaltic deposits on the surface of the ground.

Oil seeps in the Wyoming part of the western thrust belt were first reported in the 1850's by pioneers who were migrating across the United States along the Overland Trail. These seeps are in the Cretaceous Aspen Shale exposed at the surface in front of the Absaroka thrust fault, and are located about 15 miles southeast of the present city of Evanston, Wyoming (Fig. 2), where the pioneers used the tarry oil to lubricate the axles on their wagons. Several small and insignificant, shallow oil fields were developed in this vicinity around the turn of the century, all of which have been abandoned.

Between 1881 and 1900, several geologists from the newly formed U.S. Geological Survey studied and mapped the rock sequences in western Wyoming in what is now the present-day productive area of the thrust belt. Perhaps the first real breakthrough in geologic understanding of this complex structural province, however, came from the studies of A. C. Veatch (9) of the fledgling U.S. Geological Survey. Veatch recognized most of the thrust faults in this area, and defined many of the stratigraphic sequences that are used currently. The first major discovery, however, was made in the Canadian Foothills salient of the thrust belt (Fig. 1). The discovery of gas at the Turner Valley field

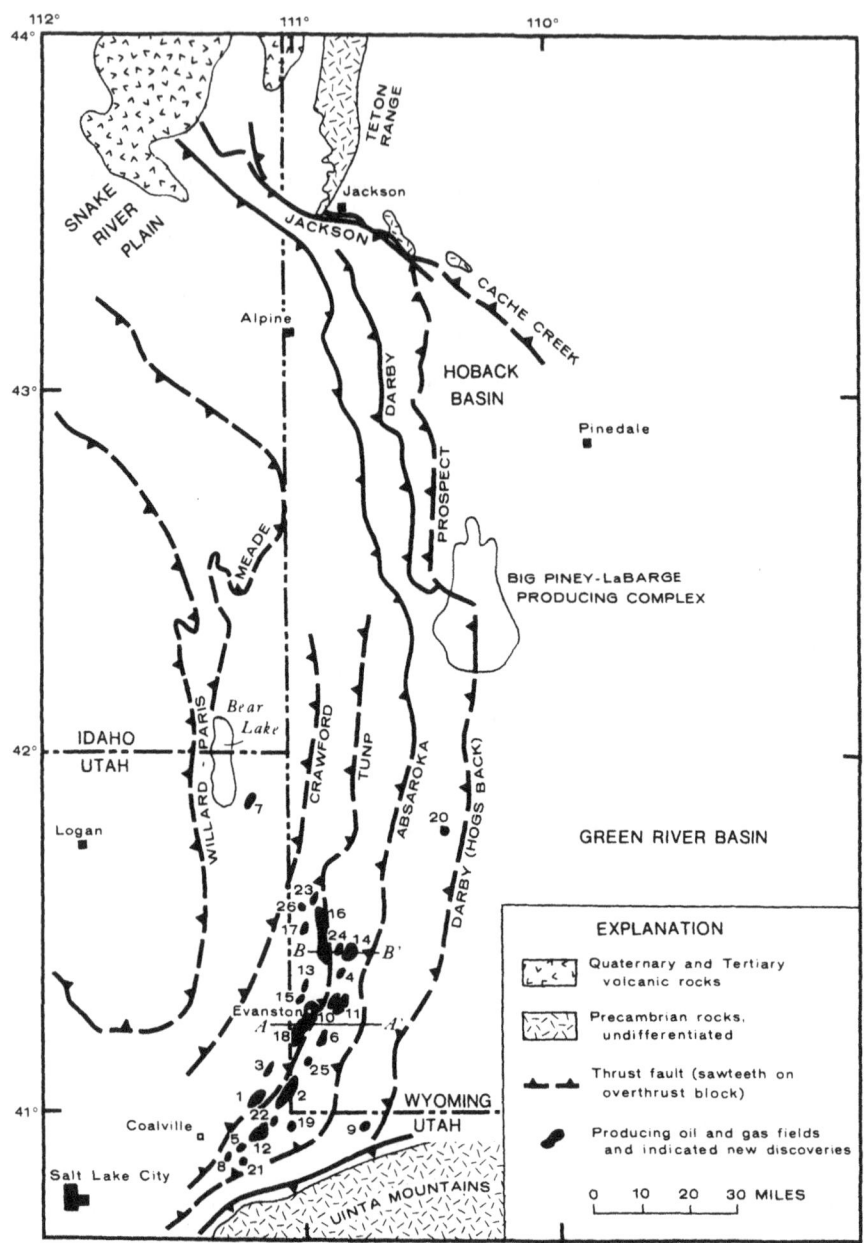

Figure 2. Index map of Wyoming-Utah-Idaho thrust belt showing principal tectonic features, major thrust faults, oil and gas fields (numbered) and new-field wildcat discovery wells since 1975. Numbers coded to field names are shown on Table I. (Modified from Powers, (3), (5); Lamerson and Royse, (6); Petroleum Information, (7); and Ver Ploeg and De Bruin, (8).

in southwestern Alberta, Canada, in 1924 (later, oil in 1936) led to the discovery and eventual development of some 275 MBO and 9.3 TCF of initially recoverable gas in the Canadian Foothills. The first major discovery in the United States' part of the thrust belt was the Pineview field discovery in northcentral Utah, in 1975.

In 1892, oil seeps of paraffin-base crude were found on the north shore of Kintla Lake in the western Montana salient of the thrust belt in what is now Glacier National Park, about 2 miles south of the Canada-United States border. In 1901, another oil seep was found north of St. Mary Lake in the eastern part of present-day Glacier National Park. This seep was eventually drilled, and by 1904 12 shallow wells were in operation in the newly designated Swift Current field, 5 of which were producing Montana's first oil in commercial quantity. This field was later abandoned, and most of the ground where these wells were located now lies at the bottom of Lake Sherburne, constructed later, near the eastern boundary of Glacier National Park (10).

It is interesting to note that the old oil seeps in southwestern Wyoming, and others like them, may actually be telling today's exploration geologists that the seeps are really indicators of a "leaking system" which contains major oil and gas accumulations (2). Studies by Royse (11) in the southwestern Wyoming area indicate that these active seeps in "footwall" (subthrust) Cretaceous rocks along the leading edge of the Absaroka thrust fault (Fig. 2) are actually generating oil and gas at the present time.

Wyoming-Utah-Idaho Salient

To bring the vast scale of the 2,400-mile-long Cordilleran thrust belt down to a size that is smaller and easier to comprehend, this discussion will concentrate on a representative fraction of the overall Cordilleran system, the Wyoming-Utah-Idaho thrust belt. As its name implies, this particular salient straddles parts of western Wyoming, northcentral Utah, and eastern Idaho (Fig. 2).

Boundaries

The total area of this salient is approximately 15,000 square miles. The northern boundary is placed at the south edge of the Snake River Volcanic Plain west of Jackson, Wyoming; the southern boundary is at the intersection of the Uinta Mountains just east of Salt Lake City. The eastern

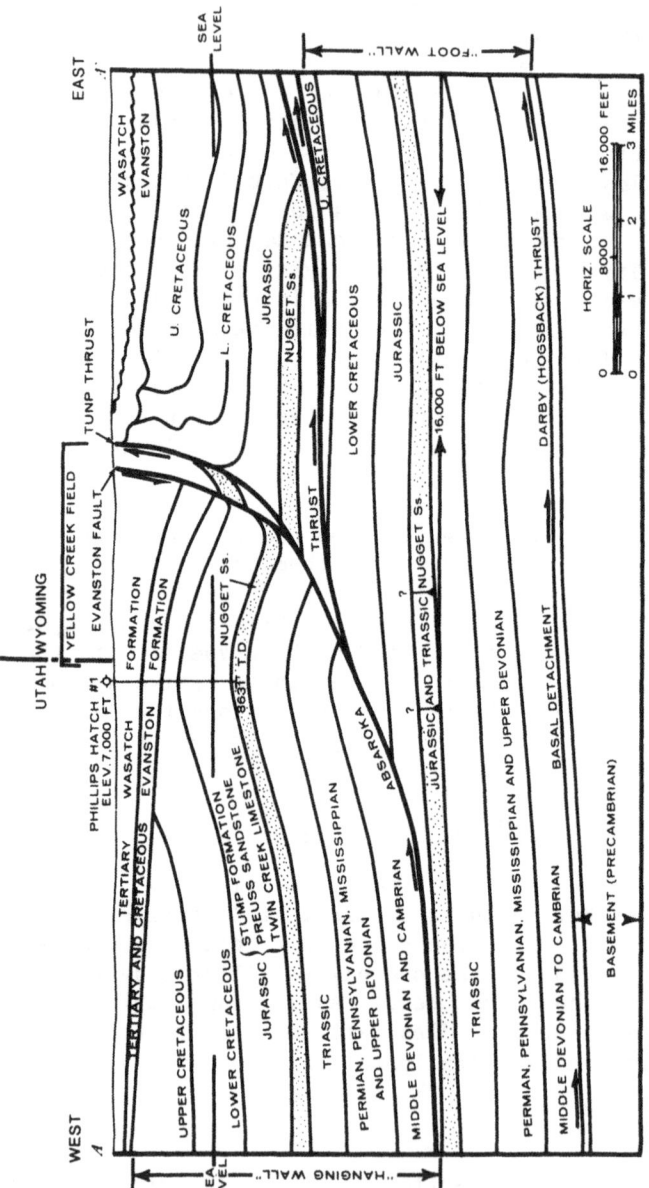

Figure 3. Diagrammatic west to east structural cross section, in vicinity of Yellow Creek field, Uinta County, Wyoming, and Summit County, Utah, showing the relation between an anticline in the "hanging wall", and rocks in the "footwall" (subthrust block). Older Devonian and Cambrian rocks have been thrust over younger Cretaceous and Jurassic rocks along the Absaroka thrust fault (basement is not involved in thrusting). (Slightly modified from Royse and others, (4), Plate II, and Conrad, (13)). Line of cross section is shown in Figure 2.

boundary is at the surface trace of the Darby-Prospect thrust fault, which separates the structurally deformed thrust belt from the undeformed Green River Basin on the east. The western boundary is the surface trace of the Willard-Paris thrust on the east side of the Bear River Range (Fig. 2). Contrary to some recent newspaper articles (12), the thrust belt does not extend through Yellowstone or Grand Teton National Parks.

Geologic Framework

During a span of geologic time (Paleozoic-early Mesozoic) ranging from 570 to 140 million years before the present (m.y. B.P.), more than 60,000 ft. of sand, silt, mud and limy material was deposited in an ocean basin that was located some distance west of the present thrust belt. About 140 m.y. ago, in latest Jurassic time, the ocean basin (or miogeosyncline) began to be deformed, and the mass of sedimentary rocks within it was strongly folded and thrust eastward by compressional forces from the west. Compression continued episodically for about 85 m.y. or until early Eocene time (55 m.y.) after which normal (extensional) faulting occurred within the thrust-fold belt until the present. Eastward horizontal movement on most of the individual thrust sheets exceeds 5 miles.

Four major thrust-fault systems are recognized in this province and are named after the major thrusts involved (Fig. 2). They are from oldest to youngest, and from west to east: (1) Willard-Paris, (2) Meade-Crawford, (3) Absaroka, and (4) Darby-Prospect-Hogsback (11). Prior to horizontal shortening of some 65 mi due to west-east folding and thrusting, the width of this rock sequence was about 130 miles. The chief structural characteristic of this province, as a result of all of the tectonic pushing, sliding, and crumpling is that older rocks have been thrust over, and now overlie younger rocks. This is illustrated in a cross section (Fig. 3), in which rocks as old as Devonian and Cambrian in the hanging wall (overthrust block) are thrust over rocks as young as Cretaceous in the footwall (sub-thrust block), forming an anticlinal fold, or possible trap for oil and gas, in the hanging wall.

Thrust movements as described above present mixed blessings to petroleum exploration geologists in that they have made interpretation of the structural geometry of these complexly folded rocks much more difficult and time consuming. The positive aspects of these complexities, however, are that the folding and thrusting have resulted in

Geologic Age			Formation or Group		Oil or Gas	Thickness Range
Tertiary			Green River Fm.			0–8,000'
			Wasatch-Evanston Fms.			
Cretaceous	Late		Adaville Fm.			6,000'– 16,000'
			Hilliard Fm.			
			Frontier Fm.		RS ●	
	Early		Aspen Shale		RS ●	10,000'
			Bear River Fm.		RS ✳	
			Gannett Group			
Jurassic			Stump Fm.		R ●	500'– 1,000'
			Preuss Ss. ⟨Salt[1]		R ✳	
			Twin Creek Ls.		RS ✳	1,200'– 3,500'
			Gypsum Spring Mbr.			
Jurassic(?) and Triassic(?)			Nugget Ss.		R ✳	500'– 2,000
Triassic			Ankareh Fm.		R ✳	2,000'– 7,000'
Early Triassic			Thaynes Fm.		RS ☼	
			Woodside Fm.			
			Dinwoody Fm.		R ☼	
Permian			Phosphoria Fm.[2]		RS ☼c	400'–5,000'
Pennsylvanian			Weber Ss.		R ☼c	750'–4,000'
Mississippian	Madison Group		Mission Canyon Ls.[3]		R ☼c	1,000'– 7,000'
			Lodgepole Ls.		RS ☼c	
Devonian		Darby FM[4]	Three Forks Fm.		RS ☼c	500'– 3,000
			Jefferson Fm.			
Ordovician			Bighorn Dolomite		RS ☼c	250'–2,000'
Cambrian			Gallatin Fm.			1,500'– 5,000'
			Gros Ventre Fm.			
			Flathead Ss.			
Precambrian			Uinta Mtn. Group			>20,000'

● Oil productive   ☼c Gas with condensate productive

✳ Oil and gas productive   R- Known or potential reservoir rock

☼ Gas productive   S- Known or potential source rock

[1]of Maher (1976)   [3]Brazer limestone of some authors

[2] And equivalent strata   [4]Locally in Wyoming

Figure 4. Generalized stratigraphic chart of the Wyoming-Utah-Idaho thrust belt showing productive formations and known, or potential reservoir and source rocks. (Modified from Hayes, (2); Powers, (5), (15); Lageson and others, (16); and Ver Ploeg and De Bruin, (8)).

the formation of hydrocarbon traps in which giant-size[1] oil and gas fields have been discovered.

When evaluating the oil and gas (hydrocarbon) potential of the thrust belt, as well as for most sedimentary basins, several factors critical to the generation, migration and trapping of hydrocarbons must be present and in natural balance, in order for discrete hydrocarbon accumulations to occur. The most important of these factors are: 1) number and thickness of porous reservoir rocks, 2) organic-rich fine-grained rocks that are the source of hydrocarbons, 3) thermal maturation of the source rocks that allows the generation and expulsion of hydrocarbons, 4) structural or stratigraphic traps such as folds (anticlines), faults, or lensing of reservoir rocks, 5) tight, dense (impermeable) rocks over the trap that act as a seal to prevent the upward and lateral escape of hydrocarbons, and finally 6) correct timing as to generation and migration of hydrocarbons relative to the forming of a trap. All of these factors are present in the Wyoming-Utah-Idaho thrust belt.

Reservoir Rocks

Fifteen formations produce oil or gas in anticlinal traps in the thrust belt. These formations represent eight different geologic systems, ranging in age from Cretaceous through Ordovician (Fig. 4). In contrast, many producing basins in the United States have only one or two formations within a single geologic period that are oil or gas bearing.

Rocks of the Jurassic and Triassic contain six reservoir formations that are productive of oil or gas, including the prolific Nugget Sandstone, which is the main oil or gas reservoir in 11 fields. Paleozoic rocks contain five productive reservoirs ranging in age from the Permian Phosphoria Formation through the Ordovician Bighorn Dolomite. These formations are the major gas and condensate (light, high gravity crude oil similar to natural gasoline) reservoirs in seven recently discovered fields in the thrust belt (Table 1, Fig. 4). The Madison Group of Mississippian age, which is the principal Paleozoic reservoir in these fields, has substantial intercrystalline porosity developed in limestones and dolomites of the Mission Canyon Limestone (Fig. 4).

---

[1]A giant field is defined as one that is expected to produce more than 100 MBO, or 1 TCF of combustible gas (14).

Table 1. New fields discovered since 1975, and indicated (one well drilled) recent new-field wildcat discoveries in the Wyoming-Utah-Idaho thrust belt. Numbered map locations, of fields are shown on Figure 2. Ages of producing formations are shown on Figure 4. (Modified and updated from Petroleum Information, (7); and Ver Ploeg and De Bruin, (8).)

Field name and number	Producing formations	Hydrocarbon type	Thrust plate	Discovery date
1. Anschutz Ranch	Twin Creek Limestone Nugget Sandstone Weber Sandstone	Sweet gas and condensate Sweet gas and condensate Sour gas and condensate	Absaroka (East)[1] Absaroka (West)	10/78
2. Anschutz Ranch East	Nugget Sandstone	Sweet gas and condensate	Absaroka (East)	12/79
3. Cave Creek	Phosphoria Formation Weber Sandstone Mission Canyon Limestone	Sour gas and condensate ----------do---------- ----------do----------	Absaroka (West)	10/79
4. Clear Creek	Nugget Sandstone	Oil and sweet gas	Absaroka (East)	8/78
5. Elkhorn Ridge	Twin Creek Limestone	Oil and sweet gas	Absaroka (East)	9/77
6. Glasscock Hollow	Nugget Sandstone	Sweet gas and condensate	Absaroka (East)	9/80
7. Hogback Ridge	Dinwoody Formation Phosphoria Formation	Dry sweet gas Sour gas	Crawford	10/77
8. Lodgepole	Twin Creek Limestone Nugget Sandstone	Oil and sweet gas ----------do----------	Absaroka (East)	3/77
9. Mill Creek (Temporarily abandoned)	Darby Formation	Oil and sweet gas	Darby-Hogsback	4/80

Table 1. Continued.

Field name and number	Producing formations	Hydrocarbon type	Thrust plate	Discovery date
10. Painter Reservoir	Nugget Sandstone	Oil and sweet gas	Absaroka (East)	10/77
11. Painter Reservoir East	Nugget Sandstone	Sweet gas and condensate	Absaroka (East)	8/79
12. Pineview	Frontier Formation Stump Formation Twin Creek Limestone Nugget Sandstone	Oil and sweet gas -------do------- -------do------- -------do-------	Absaroka (East)	1/75
13. Red Canyon	Weber Sandstone	Sour gas and condensate	Absaroka (West)	12/79
14. Ryckman Creek	Nugget Sandstone Ankareh Formation Thaynes Formation	Oil and sweet gas Sweet gas and condensate -------do-------	Absaroka (East)	9/76
15. Thomas Canyon	Phosphoria Formation Madison Group	Sour gas and condensate -------do-------	Absaroka (West)	8/81
16. Whitney Canyon– Carter Creek	Thaynes Formation Phosphoria Formation Weber Sandstone Mission Canyon Limestone Lodgepole Limestone Darby Formation Bighorn Dolomite	Sweet gas and condensate Sour gas and condensate -------do------- -------do------- -------do------- -------do------- -------do-------	Absaroka (West)	8/77

Table 1. Continued.

Field name and number	Producing formations	Hydrocarbon type	Thrust plate	Discovery date
17. Woodruff Narrows	Bighorn Dolomite	Sour gas and condensate	Absaroka (West)	4/81
18. Yellow Creek	Twin Creek Limestone Phosphoria Formation Weber Sandstone	Sweet gas and condensate Sour gas and condensate ----------do----------	Absaroka (West)	7/7b
19. [2]Aagard	Frontier Formation Stump Formation	Oil and sweet gas ----------do----------	Absaroka (East)	3/82
20. Horsetrap	Madison Group	Sweet gas and condensate	Darby-Hogsback	6/82
21. Lodgepole South	Frontier Formation	Sweet gas	Absaroka (East)	9/78
22. North Pineview	Nugget Sandstone	Sweet gas and condensate	Absaroka (East)	9/82
23. Road Hollow	Bighorn Dolomite	Sour gas and condensate	Absaroka (West)	11/81
24. Ryckman Creek West	Ankareh Formation	Oil and sweet gas	Absaroka (East)	2/82
25. Coyote Creek	Nugget Sandstone	Sweet gas and condensate	Absaroka (East)	11/82
26. West Carter Creek	Mission Canyon Limestone Bighorn Dolomite	Sour gas and condensate ----------do----------	Absaroka (West)	11/82

[1]Position of field east or west of Tunp imbricate thrust of main Absaroka thrust.
[2]Numbers 19-26 are indicated new-field wildcat discoveries.

Source and Seal Rocks

Nine formations in the thrust belt contain possible source rocks ranging from the Cretaceous Frontier Formation to the Devonian Darby Formation (Fig. 4). However, the only probable source rocks, for which documented and published data exist, are those of Cretaceous age (Frontier Formation, Aspen Shale, and Bear River Formation). Warner (17) conducted extensive geochemical studies and concluded that the oil trapped in Jurassic-Triassic reservoir rocks in fields along the Ryckman Creek-Pineview structural trend on the Absaroka thrust (Fig. 2) was generated in source rocks in the footwall Cretaceous sequence that was overriden by the Absaroka thrust plate (about 75 m.y. ago). This conclusion is supported by data from a variety of analytic techniques used to make oil-to-oil and oil-to-source rock correlations. The source of sour gas (gas high in sulfur content) and high-gravity oil in Paleozoic reservoirs in the fields along the Whitney Canyon-Carter Creek trend west of the Tunp thrust (Fig. 2) is more difficult to identify with any degree of certainty. However, the identical sulfur content and chromatographic character of condensate from the Paleozoic and Jurassic-Triassic reservoirs suggest that both were generated from the same Cretaceous source rocks (17), or possibly at the same time from rocks of different ages.

The key factors that led to the presence of oil and gas fields in the thrust belt appear to be: 1) the presence of an extensive area of organic-rich Cretaceous source rocks at peak maturation in the footwall of, and in contact with, the Absaroka thrust, 2) generation of oil and gas from these rocks after being overridden and buried by the Absaroka plate, and consequent expulsion of the oil and gas, 3) migration of the expelled hydrocarbons laterally and upward into Jurassic, Triassic and Paleozoic reservoir rocks in available hanging wall traps, and 4) sealing over the traps by impermeable shale, anhydrite, or halite (salt) caprocks. It is probable that other structural traps in the thrust belt, although highly prospective, may be barren where these four conditions are not met (11).

Structure

As stated earlier, a wedge of Paleozoic and Mesozoic sedimentary rocks was compressed from west to east into a zone about one-half of its original width, resulting in the thrust folds of the present Wyoming-Utah-Idaho thrust belt. However, in this province only the sedimentary rock section is involved in folding and thrusting. The sedimentary rock section is structurally detached from

Figure 5. West–east cross section view of Little Greys River anticline, Lincoln County, Wyoming. Middle center of photo is the axis of the anticline. Jurassic sedimentary rocks dip 45° on the east limb and 36° on the west limb. (Photo by the author.)

basement crystalline rocks (Precambrian granite) by a
regional "décollement". This décollement condition has also
been referred to as "thin-skinned" structure by Rodgers (18)
in describing folds and faults in a thrust belt involving
only the upper strata lying on a décollement, beneath which
the structure differs. Structurally, the basement is
passive and only the overlying sedimentary rock sequence is
actively involved in the shortening (Fig. 3). Nearly all of
the present oil and gas fields in the thrust belt have been
trapped in asymmetric and overturned anticlinal folds in the
hanging wall of the Absaroka thrust. Most of the folds have
numerous, additional imbricate thrust faults included within
their overall configuration. Vertical structural relief or
amplitude of the traps ranges from 500 to over 4,500 ft, and
areal size ranges from about 3 to more than 50 square
miles. The greater the vertical relief and area of the
fold, the larger the amount of oil or gas it can trap. Many
of the new discoveries in the thrust belt are major fields[1]
and some are estimated to be giant fields.

Many examples of these structural fold types are
present on the surface in Wyoming and Idaho. A near-
textbook example of such a fold is Little Greys River
anticline in Lincoln County, Wyoming, about 18 miles south
of Jackson Hole (Fig. 5). Recent U.S. Geological Survey
mapping (20)indicates that this feature is approximately 8
miles long and 2 miles wide and has an estimated 3,000 ft of
vertical structural closure. A good example of extremely
tight folding in the same general area is the St. John
thrust complex in Snake River Canyon, 2 miles east of
Alpine, Wyoming (Fig. 6). Geologic mapping by Jobin (21)
indicates that the east limb of this fold is overturned by
15° from the vertical in the Mississippian Mission Canyon
Limestone.

The upper plate of the Prospect thrust on the Hoback
River, southern Teton County, Wyoming, includes a number of
tightly folded, overturned anticlines. Some of these
features are so tightly folded that the sedimentary rocks
involved in the fold axes have a steeply dipping "chevron"
configuration (Fig. 7). An extreme example of just how
complex some thrust folds are can be observed on the
Medicine Lodge thrust. At Scott Peak, on this thrust, in

---

[1]A major field is defined informally as one that is
estimated to ultimately produce 50 MBO or more, or 300 BCF
or more of combustible gas (19).

Figure 6. St. John thrust complex, looking north, in Snake River Canyon 2 miles east of Alpine, Wyoming. The axis of this tight fold is in the center of the photo in the Devonian Darby Formation. Thrust-fault contact follows along the crest of the ridge to the right of the fold. (Photo by the author.)

Figure 7. Aerial view of sharply folded Hunter Creek anticline on the Prospect thrust, Hoback River, southern Teton County, Wyoming. Older Paleozoic rocks (hanging wall) on the left, within the fold, are resting on younger Jurassic rocks (footwall) on the right at the thrust-fault contact. View is northwest. (Courtesy of W. R. Sacrison).

Figure 8. Scott Peak segment of the Medicine Lodge thrust, an example of highly complex, multiple thrust folding. View is west-northwest with Scott Peak (elevation 13,393 ft) at top left of photo. Location is in easternmost Lemhi County, Idaho, near the Montana State line. Rocks involved in the complex are Late Mississippian in age, and vertical relief is 1,400 ft (Courtesy of W. B. Hanson).

easternmost Lehmi County, Idaho, a series of tightly
compressed anticlines are actually lying on their sides with
axial planes oriented in a horizontal position (Fig. 8).
These are referred to as recumbent folds or more commonly as
chevron or "Z" folds because of their distinct geometric
configuration.

Exploration Techniques and Drilling

Exploring for, and locating structural folds of the
types discussed above initially involves a great deal of
field mapping and measurement of the structures at the
surface. Exploration techniques used in 1953 may seem
archaic when compared to those used in 1983. Thirty years
ago direct field mapping of surface structures relied
heavily on the use of the plane-table alidade method of
plotting differences in elevation on a selected sedimentary
formation or "marker" bed, and connecting these points on a
chart (Fig. 9). The result was a structure contour map of
the configuration of a prospective surface anticline.
Transportation was slow, usually by field vehicle or
horseback, and a great deal of "foot power" was used in
moving from map point to map point (observation point).
Today, in contrast, the exploration geologist can select an
area of investigation from aerial or Earth satellite photos,
and a helicopter can quickly drop him on a specific outcrop
within that area (Fig. 10). This approach enables the
geologist to examine, map, and sample significantly more of
the surface rock outcrops in the area in one day than the
weeks or months it required previously.

The helicopter has become the work horse in today's
thrust belt exploration. The "chopper" has the unique
ability of overcoming the negative factors of extreme
variation in terrain, elevation, inaccessibility by vehicle
and by foot to surface locations, and especially in reducing
the amount of actual travel time spent on field studies. In
addition, nearly all portable seismic exploration in the
thrust belt is now done by field crews and equipment using
helicopters as the primary transportation medium. The
advantages in using helicopter methods in portable seismic
exploration include: 1) minimum disturbance to ground
vegetation, 2) access roads for tracked vehicles are not
required, 3) crews, equipment and central recording trailers
can be dropped on specific ground spots, and 4) seismic
recording cables can be layed out on a continuous, straight
line (no offsets or curves), resulting in excellent quality
of final seismic records. Of course, some of the basic
costs involved are higher than conventional seismic work.

Figure 9. Plane-table alidade method of field mapping a surface structure in the mid-1950's. Illustrates instrument man lining up and recording elevation and distance to a specific location on a marker bed identified by field partner (out of picture) who uses a 15 ft "stadia" measuring rod. View is west, with oil shales of the Tertiary Green River Formation in background. (From author's collection.)

Figure 10. Exploration helicopter approaching outcrop of Jurassic Twin Creek Limestone (above) and Jurassic and Triassic Nugget Sandstone (below) on forested slopes of Indian Peak, Lincoln County, Wyoming. View is northwest. (Courtesy of Harvey R. Duchene.)

For example, the average cost of seismic exploration utilizing surface-energy explosive charges is $12,000 to $17,000 per line-mile, and nearly twice that amount if shot-point (explosive charges are placed in drilled holes) drill rigs are used. Average cost for helicopter use is $500 to $600 per hour, plus a $25,000 fixed monthly operating fee. These high costs involved in helicopter-portable seismic exploration are more than justified, however, on the basis of the resulting discoveries of large oil and gas fields in the past 8 years in this province.

Similarly, because of the complex structure of the folds, drilling costs in the thrust belt are also much higher than in most other exploratory provinces of the United States that contain relatively flat-lying rocks that are structurally less complex and easier to drill. For example, an average wildcat well in the 14,000-15,000 ft depth range in the Whitney Canyon-Carter Creek field (Fig. 2) costs approximately $10 million to drill and place on production (W. R. Sacrison, personal commun.). Few of the holes drilled in this, or other nearby fields, are straight vertical holes; this is due to the great number of faults and steeply dipping rocks that cause the drill bit to deviate repeatedly before reaching the target oil or gas reservoir. As a result, the subsurface location of most wells is usually different from the surface location because of the deviation. This problem, among others, causes a significant increase in overall drilling costs.

Published U.S. Geological Survey maps of these folded terrains within the Wyoming-Utah-Idaho thrust belt initially provide the exploration geologist with tangible evidence as to the size, amount of structural relief, quality of reservoir and source rocks, and, most critically, the geometric configuration of the surface folds themselves. Lack of success in the past in finding oil or gas traps in most of the surface structures eventually led to the conclusion that perhaps the real structures that held oil and gas did not necessarily lie directly beneath the surface structures. The problem then became a matter of finding the right tool, or method, that would enable explorationists to look thousands of feet below the surface and identify anticlines similar to those on the surface that might trap significant amounts of oil or gas.

The "eyes" that became available to solve the problem were an improved method of subsurface mapping employing advanced seismic-reflection tools that were the result of a breakthrough in seismic data processing and mapping, coupled with modern, sophisticated computer technology. In essence, this method allowed explorationists to find new places to look, but just as importantly, they learned how to look (22).

Two of the first buried structures found by this advanced seismic-reflection technique were the Whitney Canyon, 12,600 ft deep, and Ryckman Creek fields, 7,800 ft deep. Rocks on the surface in the area where these structures are located are flat lying and give no clue whatever of the presence of these two thrust folds at depth (Fig. 11). The interpreted seismic cross section illustrated in Figure 11 shows a thick sequence of younger Tertiary rocks that are nearly horizontal lying on older, folded Jurassic and Triassic (Nugget sandstone) and Mississippian (Madison Group) rocks. Tertiary rocks are not thrusted or folded, indicating that they were deposited some time after thrusting had ended, and thus effectively masking the presence of the deeply buried structures. The illustration also shows that both the Whitney Canyon and Ryckman Creek structures are folds in the hanging wall of the Absaroka thrust, and are structurally closed (sealed) to the east against subsidiary thrust faults. The sole thrust at the base of the seismic section (Darby thrust fault) eventually rises to the surface a few miles eastward. The structural setting shown in Figure 11 is fairly typical of other settings in which new fields have been discovered in the thrust belt. Both of these structures are now giant gas and oil fields presently being developed.

Two important factors leading to the discovery of these new fields are: 1) subsurface mapping, based on advanced seismic-reflection data, that clearly identified two large anticlinal folds on the Absaroka thrust fault; and 2) exploration geologists and geophysicists believed that the folds interpreted on seismic records were real, because their geometric configurations were nearly identical to folds of similar geometry that they had observed on the surface. In other words, they had applied their knowledge of surface fold geometry directly to its subsurface counterpart and developed a working analog.

A dramatic example of geometric similarity in a thrust fold on the surface and one buried nearly 8,000 ft below the surface is illustrated in Figures 12 and 13. A thrust fold on the hanging wall of the Indian Creek thrust in Teton County, Montana, shows about 15° of overturn from vertical on the east flank of the fold (Fig. 12). Comparison of this surface feature with a west-east cross section, based on seismic mapping of the structure forming the oil and gas

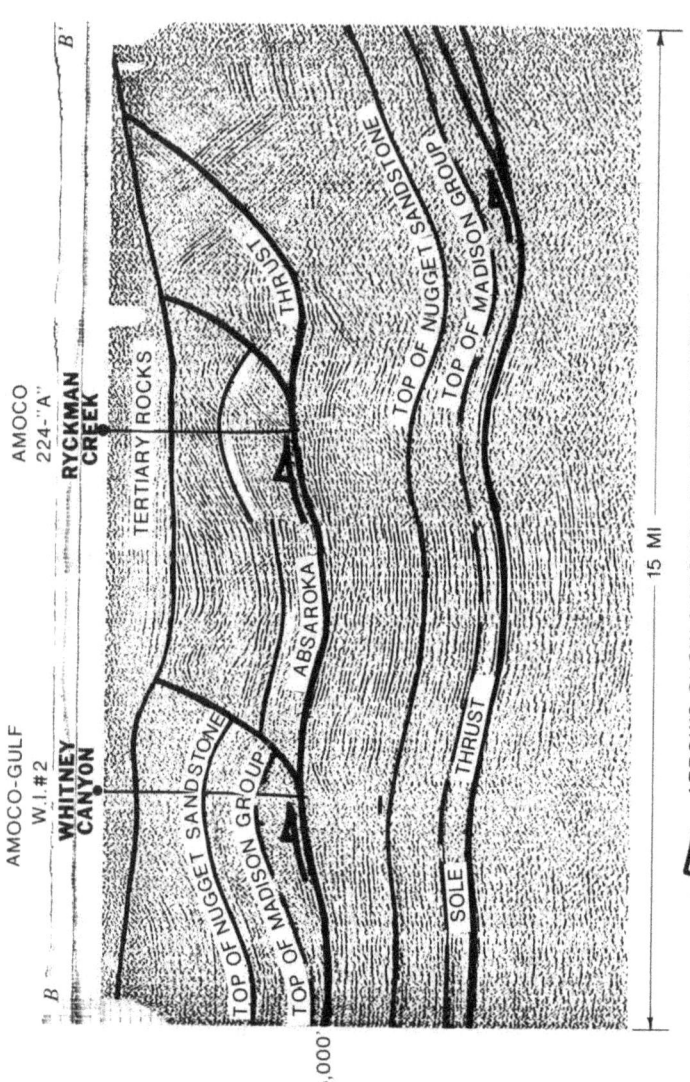

Figure 11. Generalized, compressed, west-east seismic cross section B-B' across Whitney Canyon and Ryckman Creek fields, Uinta County, Wyoming. Distance between fields is 6 miles and depth to the Absaroka thrust is 16,000 ft. This section illustrates that the two anticlines have no structural expression on the surface, and that these deeper structures were identified primarily on the basis of seismic mapping (modified from Work, (23)). Line of cross section is shown in Figure 2.

Figure 12. Surface outcrop of an overturned thrust fold, on the hanging wall of the Indian Creek thrust fault on the South Fork of the Teton River, Teton County, Montana. Photo is reversed from original view to the south for comparison with Figure 13. Rocks involved in the fold are the Jurassic Rierdon Formation. East flank is overturned 15° from vertical. Approximate scale is 60 ft horizontal and 30 ft vertical. (Courtesy of W. R. Sacrison.)

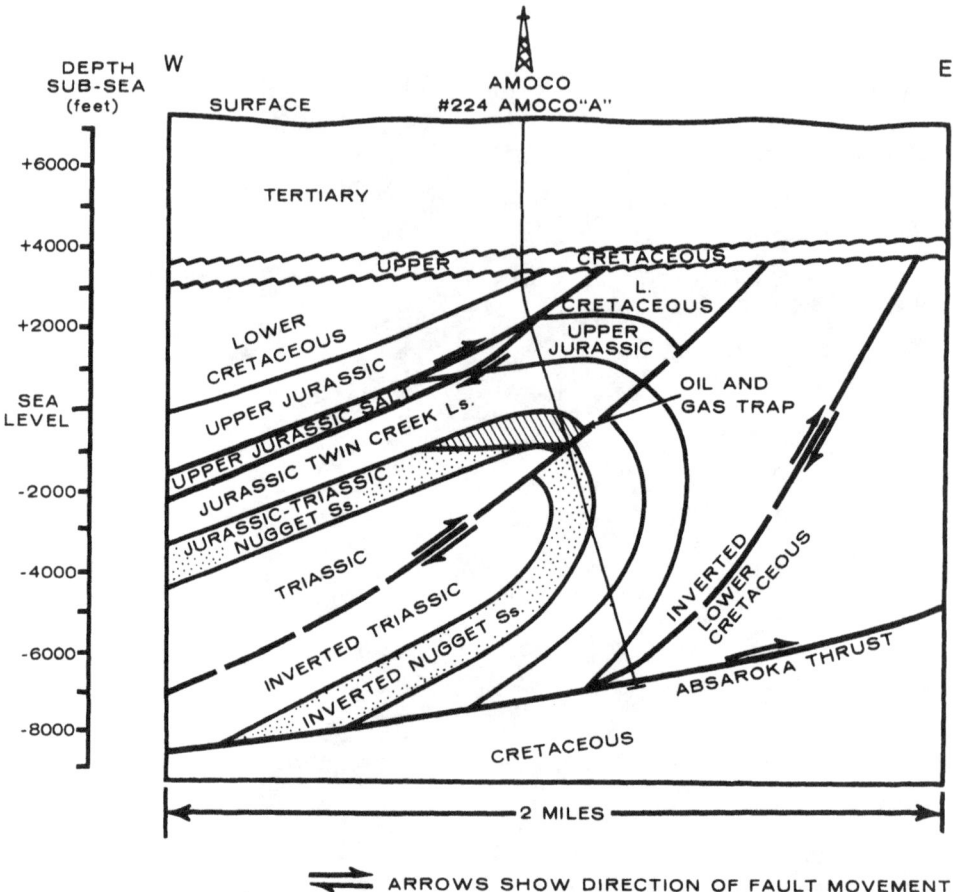

Figure 13. West-east diagrammatic cross section based on
seismic data of the structure in the Ryckman Creek field,
Uinta County, Wyoming; shows the geometric configuration of
a subsurface thrust-faulted anticline not discernable on the
surface. Note deviation of drillhole. Arrows indicate
direction of fault movement. (Modified from Powers, (5);
and Ver Ploeg and De Bruin, (8)). Line of cross section is
included in the east half of line B-B' of Figures 2 and 11.

trap in the Ryckman Creek field (Fig. 13), shows that the fold geometry of the two is essentially identical, except for actual scale, including the amount of overturn on the east flank. These same conditions are applicable to structures in the majority of new fields discovered in the province.

Typical Oil and Gas Fields

The largest field discovered to date in the thrust belt is the Whitney Canyon-Carter Creek gas field in Uinta and Lincoln Counties, Wyoming, located 13 miles north of the town of Evanston (Fig. 2, Table 1). The field was discovered in late 1977 on a major north-south-trending, slightly overturned anticline on the hanging wall of the Absaroka thrust plate. The overall anticline is actually made up of three individual structures with total structural closure (vertical relief) exceeding 4,500 ft (Fig. 14). The anticline is 16 miles long and 4 miles wide (60 mi^2), or about one and one-half townships.

An unusual feature of this field is that production of sour gas, averaging 12 percent hydrogen sulfide (H_2S), and condensate comes from 6 separate reservoirs, including the Permian Phosphoria Formation, Pennsylvanian Weber Sandstone, Mississippian Mission Canyon and Lodgepole Limestones, Devonian Darby Formation and Ordovician Bighorn Dolomite (Figs. 4 and 15). A seventh reservoir, the Triassic Thaynes Formation, which is productive of sweet gas and condensate, is the exception in this dominantly sour gas field. Depth to production ranges from 9,200 ft to 14,000 ft in these formations. Warping (folding) of the upper, main Absaroka thrust sheet has evidently been the main reason for the great amount of structural closure on the overall anticline.

To realize the productive capability of this giant field, one well, the Champlin-Amoco No. 457-A (Fig. 15), was flow-tested in the Paleozoic reservoirs for a total recovery rate of nearly 75 million cubic feet (MMCF) of gas and 1,294 barrels of condensate (B.C.) per day; even more significant is the fact that a 30-ft-thick section of the Mission Canyon Limestone reservoir, alone, flowed 32 MMCF of gas per day. The total flow rate of 75 MMCF per day of gas converts to the energy equivalent of nearly 13,000 barrels of oil (23). It is estimated that the Whitney Canyon-Carter Creek field will ultimately produce as much as 7.5 trillion cubic feet of gas, more than 150 million barrels of condensate and natural gas liquids (NGL), and a significant amount of raw sulfur.

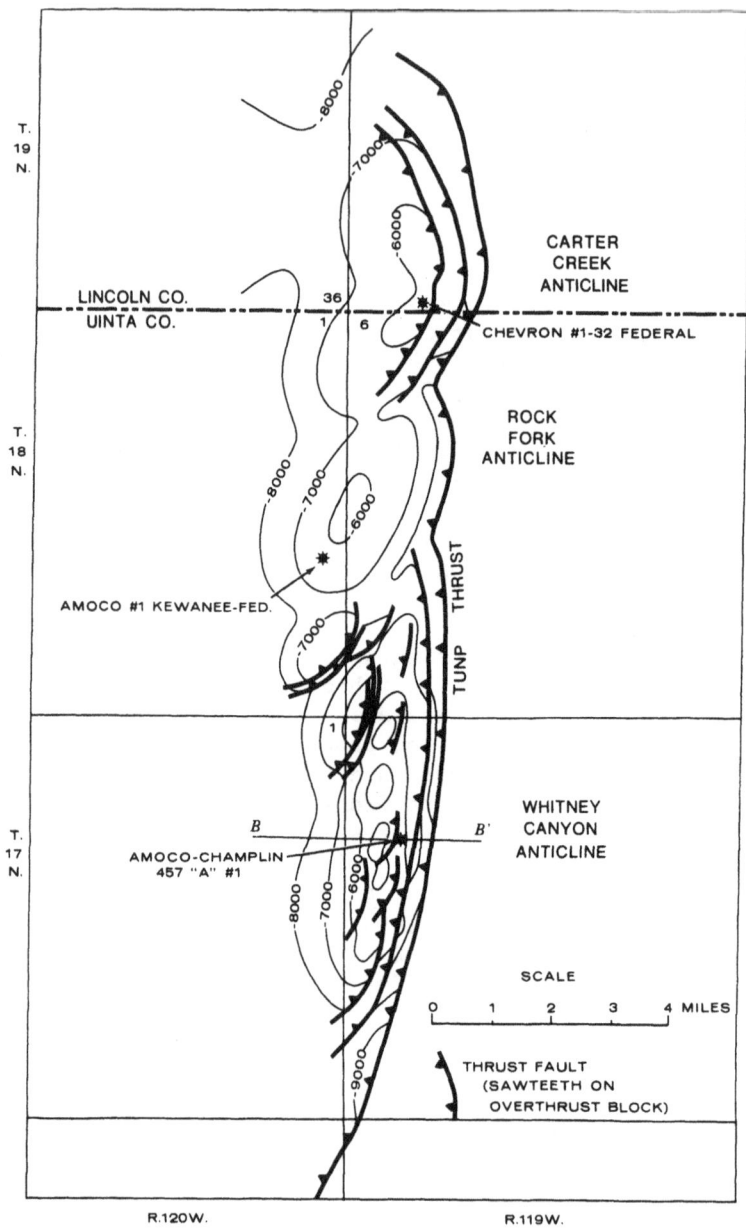

Figure 14. Structure contour map of the Whitney Canyon-
Carter Creek field, Wyoming. Contours are on top of the
Mississippian Mission Canyon Limestone; contour interval is
1,000 ft. Contours show the three large anticlines that
make up the field. Also shown are field discovery wells on
each of the three anticlines. (Modified from Hoffman and
Balcells-Baldwin, (24).)

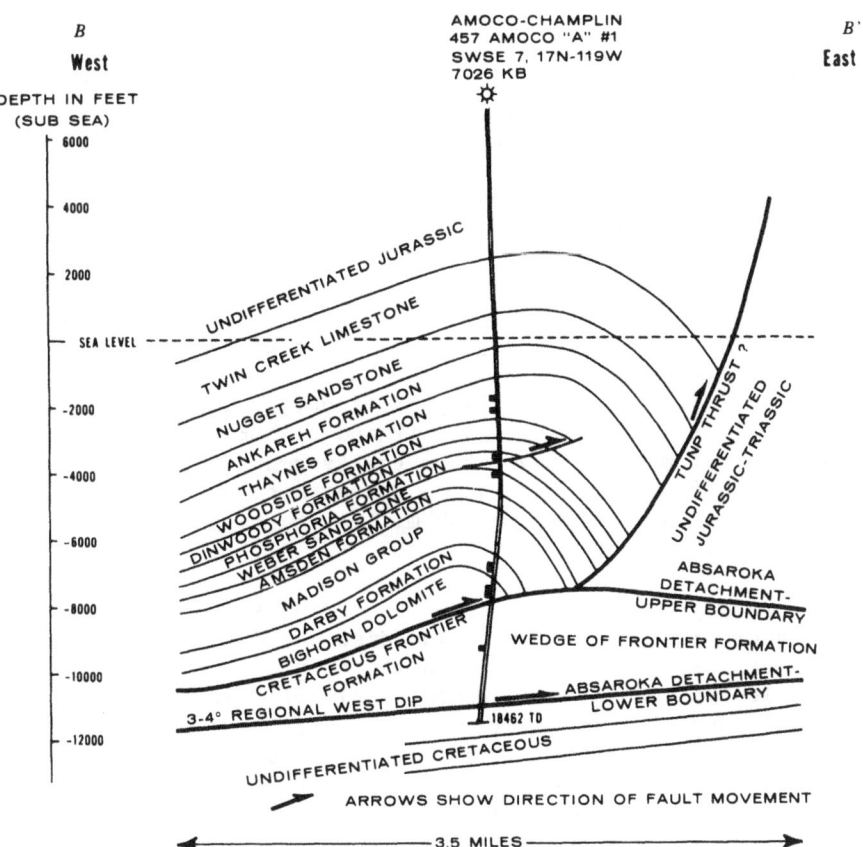

Figure 15. Simplified west-east structural cross section across Whitney Canyon-Carter Creek field in the Wyoming-Utah-Idaho thrust belt, showing hanging wall fold closed against the Tunp thrust. Also shows "warping" (folding) of the main Absaroka thrust plate. Black rectangles indicate producing intervals. (Courtesy of Amoco Production Co.). Line of section is shown in Figure 14.

One of the first fields that found oil production in the Jurassic and Triassic Nugget Sandstone in the thrust belt was at Ryckman Creek in northern Uinta County, Wyoming, 6 miles east of Whitney Canyon-Carter Creek field (Figs. 2 and 13). The field was discovered in late 1976. It was defined by seismic work as a northeast-trending overturned anticline on the hanging wall of the Absaroka thrust plate. Structural closure on this feature exceeds 2,000 ft and areal size is about 2.5 square miles. The total thickness of the Nugget Sandstone in the field is 815 ft, of which 515 ft is included in the gas/oil column (a 300-ft gas cap above a 215-ft oil column). The gas in the gas cap is rich in condensate and "sweet" (no H_2S) in contrast to the sour gas at Whitney Canyon-Carter Creek.

The initial discovery well in the field flowed 358 barrels of 47° A.P.I. gravity oil and 658 MCF of associated gas per day from the Nugget, and 1.7 MMCF of gas and 66 barrels of condensate per day from the Triassic Thaynes Formation (7). Average daily production, in late 1982, from the field was 4,166 barrels of oil and 31 MCF of gas (25). Estimated ultimate recovery from the Ryckman Creek Field is 50 MBO and 150 BCF of gas, which converts to a total energy equivalent of approximately 100 million barrels of oil--a giant field.

A fairly well defined, arcuate northeast-southwest pattern or trend exists of fields discovered in this part of the thrust belt (Fig. 2). In addition, within this trend, a separation occurs between predominantly oil and sweet gas fields in Jurassic and Triassic reservoirs east of the Tunp thrust, and dominantly sour gas and condensate fields in Paleozoic reservoirs west of the Tunp thrust (Fig. 2, Table 1). However, all these fields are common to the Absaroka thrust trend. Only one gas field, Hogback Ridge, is located on the Crawford-Meade thrust trend and only one recently discovered sweet gas field in a Mississippian reservoir (Horsetrap field), and four recent Mississippian sour gas discoveries (not listed) are located on the Darby-Hogsback thrust trend (Fig. 2, Table 1). Although temporarily abandoned at present, the Mill Creek field is also located on the Darby-Hogsback trend. This two-well field seems to be the exception to all of the fairly well defined patterns discussed above in that the discovery well flowed 154 barrels of sweet, 46° A.P.I. gravity green oil from a Paleozoic sandstone reservoir (Devonian Three Forks Formation). Interruptions in the established pattern of fields, trends, hydrocarbon type, and age of reservoir as shown by the anomalous Mill Creek field may indicate the possibility of a new and totally different oil and gas field trend on the Darby-Hogsback thrust. One must wonder then, how much additional undiscovered oil and gas may be found along these three developing thrust trends, assuming that similar structural and stratigraphic relationships remain constant.

Production, Pipelines and Markets

To move the various hydrocarbon products from fields in the thrust belt to the consumer requires, among other things: 1) field production facilities, 2) gas-processing plants, and 3) new and existing gas pipelines to transport the gas and other products. The Whitney Canyon-Carter Creek field, for example, presently has 17 wells connected to field production facilities. The recently completed Amoco Production Company gas-processing plant north of Evanston, Wyoming, will process the gas from these wells. This "sweetening" plant has a capacity of processing 270 million cubic feet of natural gas per day, and will remove such impurities as H_2S and water, recover and stabilize petroleum liquids, strip NGL from the gas, and convert the H_2S to inert sulfur. From this, the daily yield will be 200 million cubic feet of sweet, dry natural gas, 6,000 barrels of condensate, 12,600 barrels of NGL, and 1,200 long tons of sulfur (25).

Condensate will be transported to the Amoco Whiting Refinery near Chicago, Ill., and the NGL will be pipelined directly to Texas Gulf Coast refineries for use as primary petrochemical feedstock. Sulfur will be trucked to Kemmerer, Wyoming, 35 miles northeast of Evanston, where it will be loaded on trains for shipment to Florida and other points for use in making fertilizer and sulfuric acid.

Sweet natural gas from the plant will be fed directly into the newly completed Trailblazer Pipeline. The 36-inch-diameter pipeline, some 800 miles long, will carry the gas eastward to Beatrice, Nebraska, where the pipeline will connect to existing pipeline systems. These systems will finally carry the gas to consumer markets in the Upper Midwest and Northeast regions of the United States. The rated capacity of the Trailblazer Pipeline is approximately 525 MMCF of gas per day. The implications of this new pipeline to the nation are great, including helping to achieve the national goal of energy independence. The Commissioner of the Federal Energy Regulatory Commission recently stated, "Although this country is currently experiencing a surplus of natural gas, that surplus may be temporary. The energy shortages of just a few short years ago may reappear unless we maintain our efforts for

Table 2. Estimates of undiscovered recoverable oil and gas resources in the Wyoming-Utah-Idaho thrust belt compared to total United States estimates (26).

	Crude Oil (Billion Barrels)			Total Gas (Trillion Cubic Feet)		
	Low	High		Low	High	
	$(F_{95})^1$	$(F_5)^1$	Mean	$(F_{95})^1$	$(F_5)^1$	Mean
Wyoming-Utah-Idaho thrust belt	2.7	13.3	6.7	29.06	105.38	58.43
Total United States	64.3	105.1	82.6	474.6	739.3	593.8

[1] F_{95} denotes the 95th fractile; the probability (19 in 20 chances) of more than that amount.

F_5 denotes the 5th fractile; the probability (1 in 20 chances) of more than that amount.

increased energy self-sufficiency. The Trailblazer Project
will be a major contribution to that effort." (25).

Estimates of Undiscovered Oil and Gas

The Wyoming-Utah-Idaho thrust belt was only one of a
total of 137 geologic provinces evaluated for its
undiscovered oil and gas potential in a national study by
the U.S. Geological Survey (26). Another, recent study
focused on the impact of just two of these provinces on
total United States estimates of undiscovered recoverable
oil and gas resources (27). The two provinces involved in
this study were the Wyoming-Utah-Idaho thrust belt (onshore)
and the Beaufort Sea, Alaska (offshore). These authors
estimated that 18 percent of the undiscovered oil and 16
percent of the undiscovered gas in the United States lie
within these two provinces, indicating that these two
frontier exploration areas have an unusually high hydrocarbon "richness" factor (27). The thrust belt alone (based
on the mean) is estimated to contain 8 percent of the undiscovered oil and 10 percent of the undiscovered gas in the
United States. Table 2 shows the oil and gas resource
estimates of the thrust belt relative to total United States
estimates of these resources.

Any number of estimates and outright guesses have been
published as to the volume of oil and gas found (discovered)
in the new fields, and in indicated new-field discoveries in
the thrust belt; some of the estimates are extremely
optimistic, and others are conservative. However, on the
basis of available data at the present stage of development,
and by extrapolation of probable field sizes, the fields and
indicated new fields in the thrust belt have already
discovered an estimated 3.2 billion barrels of oil and other
petroleum liquids, and at least 16.5 trillion cubic feet of
gas.

Summary

What was at one time considered unlikely to ever become
a productive oil and gas province, the Wyoming-Utah-Idaho
thrust belt (and the entire United States western thrust
belt) is presently the country's most promising onshore
exploration frontier. After some 90 years of intermittent
drilling, using the exploration tools and concepts available
at the time, many explorationists simply walked away from,
or wrote off, the thrust belt because of discouraging
results (22). Exploration interest in this geologically
complex trend was rekindled in 1975, when the Pineview field
was discovered, the first giant field in the trend.

Previously, interpretation of the structural geometry of the complexly folded rocks of the thrust belt was extremely difficult. New sophisticated exploration techniques were necessary to find the hidden, anticlinal structures. Development of new, previously untried interpretive approaches in exploring for these complex fold traps can be tied to the breakthrough in seismic data processing and mapping, combined with advanced computer technology. These advances gave exploration geologists and geophysicists a reliable tool, or a set of "eyes", with which to probe the highly deformed subsurface area of the thrust belt.

Most of the known oil and gas reserves in the world are not distributed evenly over the Earth, but are concentrated in discrete areas, such as the thrust belt. Over 80 percent of the entire world's discovered oil and gas is contained in giant fields (28). Equally important is that these giant fields account for as much as 70 percent of the world's daily oil and gas production. This evidence suggests that the discovery of an anomalously high number of giant fields in this single province might well have the most significant impact since the discovery of the Prudhoe Bay field in Alaska on the future availability of adequate supplies of oil and gas for the United States.

References

1. Vanderbeek, J. W., 1981, Amoco Production Company press release for American Association of Petroleum Geologists meeting, San Francisco, California, June 2, 1981, 7 p., 3 attachments.
2. Hayes, K. H., 1976, A discussion of the geology of the southeastern Canadian Cordillera and its comparison to the Idaho-Wyoming-Utah fold and thrust belt, in Hill, J. G., ed., Rocky Mountain Association of Geologists Symposium on Geology of the Cordilleran Hingeline, p. 59-82.
3. Powers, R. B., 1980, Oil and gas potential of the Wyoming-Utah-Idaho overthrust belt in relation to its Canadian Foothills thrust belt analog (abstract): American Association of Petroleum Geologists Bulletin, v. 64, no. 5, p. 767.
4. Royse, Frank, Jr., Warner, M. A., and Reese, D. L., 1975, Thrust belt structural geometry and related stratigraphic problems, Wyoming-Idaho-northern Utah, in Bolyard, D. W., ed., Rocky Mountain Association of Geologists Symposium on deep drilling frontiers in the central Rocky Mountains, p. 41-54.

5. Powers, R. B., 1977, Assessment of oil and gas resources in the Idaho-Wyoming Thrust Belt, in Wyoming Geological Association Guidebook, 29th Annual Field Conference, Rocky Mountain thrust belt geology and resources, p. 629-637.

6. Lamerson, P. R., and Royse, Frank, Jr., 1980, Thrust belt structures, Part B, in Wyoming-Utah overthrust belt structural style, field guide to American Association of Petroleum Geologists, Trip No. 5, June 1980, Denver, Colorado, 34 p. plus maps and photos.

7. Petroleum Information Corporation, Denver, Colorado, 1981, The Overthrust Belt, 251 p.

8. Ver Ploeg, A. J., and DeBruin, R. H., 1982, The search for oil and gas in the Idaho-Wyoming-Utah salient of the overthrust belt; Geological Survey of Wyoming, Report of Investigations no. 21, 108 p.

9. Veatch, A. C., 1907, Geography and geology of a portion of southwestern Wyoming: U.S. Geological Survey Professional Paper 56, 178 p.

10. Darrow, George, 1955, The history of oil exploration in northwestern Montana 1892-1950, in Lewis, P. J., ed., Billings Geological Society Guidebook, 6th Annual Field Conference, Sweetgrass Arch and Disturbed Belt, Montana; p. 225-232.

11. Royse, Frank, Jr., 1979, Structural geology of the western Wyoming-northern Utah thrust belt and its relation to oil and gas; American Association of Petroleum Geologists Distinguished Lecture; Oil and Gas Journal, Feb. 12, v. 77, no. 7, p. 155-156.

12. Schmidt, W. E., 1981, U.S. considers drilling leases in a wilderness: New York Times, Aug. 30. 1981, p. 1 and p. 50.

13. Conrad, J. F., 1977, Surface expression of concentric folds and apparent detachment zone (décollement) near Cokeville, Wyoming, in Wyoming Geological Association Guidebook Symposium, 29th Annual Field Conference, Rocky Mountain Thrust Belt Geology and Resources, p. 385-390.

14. Halbouty, M. T., ed., 1980, Giant oil and gas fields of the decade 1968-1978; American Association of Petroleum Geologists Memoir 30, 596 p.

15. Powers, R. B., 1983, Geologic framework and petroleum potential of the greater Bob Marshall Wilderness Area, western Montana thrust belt: U.S. Geological Survey open-file report (in press).

16. Lageson, D. R., Lowell, J. D., Lamerson, P. R., and Royse, Frank, Jr., 1980, Wyoming-Utah Overthrust Belt structural style, field guide to American Association of Petroleum Geologists Trip No. 5,

June 1980: Rocky Mountain Association of Geologists, Denver, Colorado, 34 p. plus maps and photos.

17. Warner, M. A., 1982, Source and time of generation of hydrocarbons in Fossil Basin, western Wyoming thrust belt: in Powers, R. B., ed., Geologic studies of the Cordilleran thrust belt, Rocky Mountain Association of Geologists, v. 2, p. 805-816.

18. Rodgers, John, 1963, Mechanics of Appalachian foreland folding in Pennsylvania and West Virginia: American Association of Petroleum Geologists Bulletin, v. 47, no. 8, p. 1527-1536.

19. Johnston, R. R., 1980, North American drilling activity in 1979: American Association of Petroleum Geologists Bulletin, v. 64, no. 9, p. 1295-1330.

20. Schroeder, M. L., Albee, H. F., and Lunceford, R. A., 1981, Geologic map of the Pine Creek quadrangle, Lincoln County, Wyoming, U.S. Geological Survey Map GQ-1549.

21. Jobin, D. A., 1972, Geologic map of the Ferry Peak quadrangle, Lincoln County, Wyoming, U.S. Geological Survey Map GQ-1027.

22. U.S. Geological Survey, 1982, New concepts result in finding oil and gas in unlikely places; press release by Public Information Office, U.S. Geological Survey, Central Region, Feb. 18, 1982, 2 p.

23. Work, D. F., 1980, The Overthrust Belt--yesterday a driller's graveyard, today the most significant play since Prudhoe Bay, in Landwher, M. L., ed., Exploration and economics of the petroleum industry; Southwestern Legal Foundation, v. 18, Matthew Bender and Company, p. 77-97.

24. Hoffman, M. E., and Balcells-Baldwin, R. N., 1982, Gas giant of the Wyoming thrust belt: Whitney Canyon-Carter Creek field; in Powers, R. B., ed., Geologic studies of the Cordilleran thrust belt, Rocky Mountain Association of Geologists, v. 2, p. 613-618.

25. Overthrust News, 1982, Trailblazer Pipeline System dedicated, Overthrust Industrial Association, Denver, Colorado, Issue no. 6, 20 p.

26. Dolton, G. L., Carlson, K. H., Charpentier, R. R., Coury, A. B., Crovelli, R. A., Frezon, S. E., Khan, A. S., Lister, J. H., McMullin, R. H., Pike, R. S., Powers, R. B., Scott, E. W., and Varnes, K. L., 1981, Estimates of undiscovered recoverable conventional resources of oil and gas in the United States: U.S. Geological Survey Circular 860, 87 p.

27. Coury, A. B., and Powers, R. B., 1983, Impact of the
       Wyoming-Utah-Idaho thrust belt and Beaufort Sea on
       U.S. oil and gas resource estimates: Oil and Gas
       Journal, Tulsa, Oklahoma, v. 81, no. 24 (June 13,
       1983), p. 143-152.
28. St. John, Bill, 1981, Sedimentary basins of the world
       and giant hydrocarbon accumulations: Wyoming
       Geological Association Newsletter, October, 1981;
       (abstract) p. 6.

James H. Gary, Arthur E. Lewis

8. Oil Shale: A Resource Whose Time Has Come?

Since World War I, there have been several occasions when the time seemed ripe for commercial production of shale oil from the western oil shale deposits of the United States. Each time operations were about to begin, however, new finds of conventional crude oil reserves, decreases in crude oil prices, or decreases in demand have caused operations to be discontinued.

Once again, with the large increases in crude oil prices dictated by the OPEC countries and the decline in the rate of finding new oil, 1980 seemed to be the time commercialization of shale oil production was at last to be on its way. Several companies began construction of demonstration modules of commercial size operations. Cathedral Bluffs Oil Shale Corporation (owned by Occidental Oil and Tenneco) began construction of mining facilities which by the late 1980's would result in a Modified In Situ (MIS) facility to produce about 54,000 barrels of shale oil retorted underground and an additional 40,000 barrels per day produced in above ground retorts. The shale mined and retorted above ground would provide void volume for the MIS operation. Also, Colony Oil Shale Company (owned by EXXON and TOSCO) began construction on mining and road facilities and plant site preparation for the surface retorts and supporting processing facilities to produce approximately 47,000 barrels per day of shale oil by underground mining and surface retorting. In addition construction was started on the new town of Battlement Mesa to house workers for this project. Union Oil Company of California started construction of the mining facilities and site preparation for a 10,000 barrel per day surface retorting facility. Multi-Mineral Corporation announced plans to produce nahcolite, dawsonite, and shale oil from a facility leased from the U.S. Bureau of Mines.

During this same period, PARAHO and Geokinetics had active operations in Utah and Chevron had announced its intentions of building a pilot plant scale fluidized-bed retort in their Salt Lake City refinery to provide data for the design of a commercial operation to produce shale oil in the 1990's.

In early 1981, construction on the Colorado operations was in full swing but the combination of high interest rates, double digit inflation, reduction in oil demand, and a surplus of world crude oil raised concerns with respect to the future viability of commercial shale oil production. In late November, Multi-Minerals Corporation announced it was stopping its development operations and laying off many of its employees. This was followed in December by Cathedral Bluffs Oil Shale announcing it was stopping most of its development operations to make a reassessment of the economic feasibility of proceeding.

On May 2, 1982, Exxon abruptly announced it was stopping work on the Colony Project. Tosco elected its option of selling its interests to Exxon and the only full-sized commercial-scale project underway went into a close-down mode. This left Union Oil Company of California's commercial-size module plant in Colorado and a renewed effort by White River Shale Oil Corporation in Utah as the only operations of any size still under construction.

The reasons given in all cases of work stoppage were similar and included escalating costs, high interest rates, and decline in world crude oil prices.

This raised the question as to whether or not the budding industry was beginning a rerun of the former scenarios of hope and frustration that have been the history of oil shale in the West.

In view of these happenings, it is worthwhile to explore the case for the continued commercial development of oil shale and to review the present state of the technology. The news media have publicized well the reasons given by corporations for delaying commercialization which include; the softening of world crude oil prices, the 1982 surplus of crude oil on world markets, decreases in oil consumption by the industrialized nations, world-wide recession, and high interest rates. As a contrast, little attention has been given to the reasons for continuing with the commercial development of shale oil production.

There are fairly obvious points on which it is difficult to place dollar values. The crude oil surplus is predi. .ed to be a short time aberration brought about by Saudi Arabia flooding the market to establish their dominance of the OPEC control of world prices. Their hope is to control prices at levels which will insure long term markets for their oil and minimize switching to alternate fuels. On several occasions Sheik Yamani, Saudi Arabia's Minister of Petroleum, has stated his objective is to set the price slightly lower than the price of producing liquid fuels from alternate sources such as coal and oil shale. It may be more than a coincidence that since 1975 the prices set for Saudi Arabian crude oils have tracked very closely the prices estimated by independent engineering firms for producing shale oil from Colorado oil shales.

Although the crude oil usage by the industrialized nations is predicted to decline, usage by the developing nations is predicted to increase at a higher rate and, over the long run, will more than make up for the industrialized nation's decrease. A point to remember is that consumption is expected to be limited by production by 1990 and there will once again be a strong seller's market. To get the proper perspective, estimated shale oil prices should be compared with crude oil prices at the time shale oil will be available. Construction of synthetic liquid fuel facilities requires long lead times. Facilities that were under construction, such as the Colony operation, were not expected to be at their design level of production until the 1985-87 period and their shale oil would have been competing at 1985-87 world crude oil prices. At a conservative 5% rate of increase, Saudi Arabian crude oil, selling as of this writing for $34.00 per barrel, could be selling for $43.40 in 1986. With this as a basis, consider the economic reasons for continuing with oil shale development.

The only valid economic comparison of shale oil viability is with crude oil of an equal quality. For our purposes, the best comparison is between average low-sulfur U.S. crude and syncrude from oil shale. Syncrude is a low-sulfur, low-nitrogen, and low-metals product of hydrotreating raw shale oil.

In the May, 1981, issue of World Oil, R.E. Megill addressed the complex issue of the cost of finding new oil (1). Based upon his figures the 1978 cost of finding a new oil-equivalent barrel of oil and gas in the United States is approximately $7.50. Developing these new finds (drilling production wells and installing collecting pipelines) adds an equal amount to the costs. In order to provide one barrel of

oil per day for 20 years, it is necessary to find 7300 barrels of new oil. At a cost of $15.00 per barrel per day for
20 years, even if all of this could be expensed rather than
capitalized, it means at a 50% tax rate it will require that
$55,000 be taken from profits, or borrowed, for each barrel
per day for 20 years. This compares with estimates of
$40,000 to $60,000 per barrel per day for 20 years capital
investment for shale oil production. Typically, shale oil
projects are predicted to produce at the design rate for at
least 30 years. If put on a 30 year basis, the front end
cost for a barrel per day of new conventional U.S. crude oil
will be about $165,000 ($82,500 after taxes). By either of
these comparisons, front end money for shale oil production
compares very favorably with investing in exploration for new
oil in the United States.

A strong advantage for shale oil is that the location of
oil shale is known and the quantities in place are well established. As a result there is no finding cost for shale
oil.

Dr. John D. Haun, while President of the American Association of Petroleum Geologists, estimated the cost of finding and developing a new barrel of oil in the United States
in the 1980's at about $50 per barrel finding costs and about
$50 per barrel for developing for a total cost of $100 per
barrel (2). This value is predicted for approximately 1986-
1988 and assumes exploration and development costs will continue to increase at the rate they have since 1975. Assuming
his estimates to be correct, the 1986 front end money for a
new barrel of oil per day for 20 years will be $730,000 and
even if it all can be expensed, it will still mean a net cost
to the company of about $365,000 per barrel per day. This is
the oil with which the shale oil plants now under construction will be competing.

Federal tax regulations complicate making a direct comparison between the costs of shale oil and crude oil. Some
of the costs of finding and developing a new barrel of crude
oil can be expensed while others are capitalized. Although
both kinds of dollars must be paid out on the front end,
depreciation and return on investment is not included in the
expensed portion of the cost of a produced barrel of crude
oil. Depreciation and return on investment account for more
than half the cost of a barrel of shale oil. In the long
term this is an advantage for shale oil because once the
investment is made, and the plant is in operation, it becomes
a fixed cost and the actual cost of producing a barrel of
shale oil will increase at a much smaller rate than the rate
of inflation (3).

The cost figures used above include on-site hydrogen processing facilities to produce a high quality synthetic crude oil (syncrude) with low sulfur, nitrogen, and metals contents and very little heavy ends. Syncrude can be processed in conventional U.S. refineries to produce high yields of diesel and jet fuels and home heating oils with no significant environmental impacts.

The direct operating costs for a barrel of shale oil are estimated to be equal to the operating costs of recovering a barrel of crude oil by secondary recovery methods. Operating costs plus depreciation and return on investment are comparable to the costs of producing a barrel of crude oil by tertiary recovery methods. Therefore, by economic standards of comparison, shale oil compares very favorably with new crude oil. The big unknown is the risk involved in producing 50,000 barrels per day of shale oil because to date it has not been done. Until a 50,000 barrel per day plant has been built and operated long enough to provide cost data it will be impossible to provide a realistic risk factor.

The Resource

The extent and concentration of the resource and the distribution of ownership are important factors preventing the development of oil shale. Figure 1 (4, 5) shows the location of the Piceance Creek Basin. Figure 2 illustrates the important areas underlain by the Green River formation, a geologic sedimentary formation that includes all the oil shale. The property in private ownership is shown. For the most part, this property is owned by oil companies. Figure 3 shows the part of the resource that contains 20 gal/ton or more in thicknesses of 400, 1200, or 2000 feet or more (6). This area contains approximately 720 billion barrels of oil in place, in thicknesses greater than 400 feet and richer than 20 gal.ton (7). The overburden thickness is shown in Figure 4 (6). This resource covers an area of approximately 600 square miles with none of it more than 15 miles from a central point. This is the oil shale resource of principal economic interest now and probably for the next century or more. Note that almost all this thick, rich shale is in government hands, the principal exception being the two federal leases (C-a and C-b) owned, respectively, by Rio Blanco Oil Shale Company (Gulf and Amoco) and by Cathedral Bluffs Oil Shale Company (Occidental and Tenneco).

The oil shale of economic interest in the area of private ownership consists of a thin high-grade layer of shale (Mahogany Zone), generally 30 to 150 feet in thickness (8) (Figure 5). This layer of oil shale, averaging 30 gal/ton,

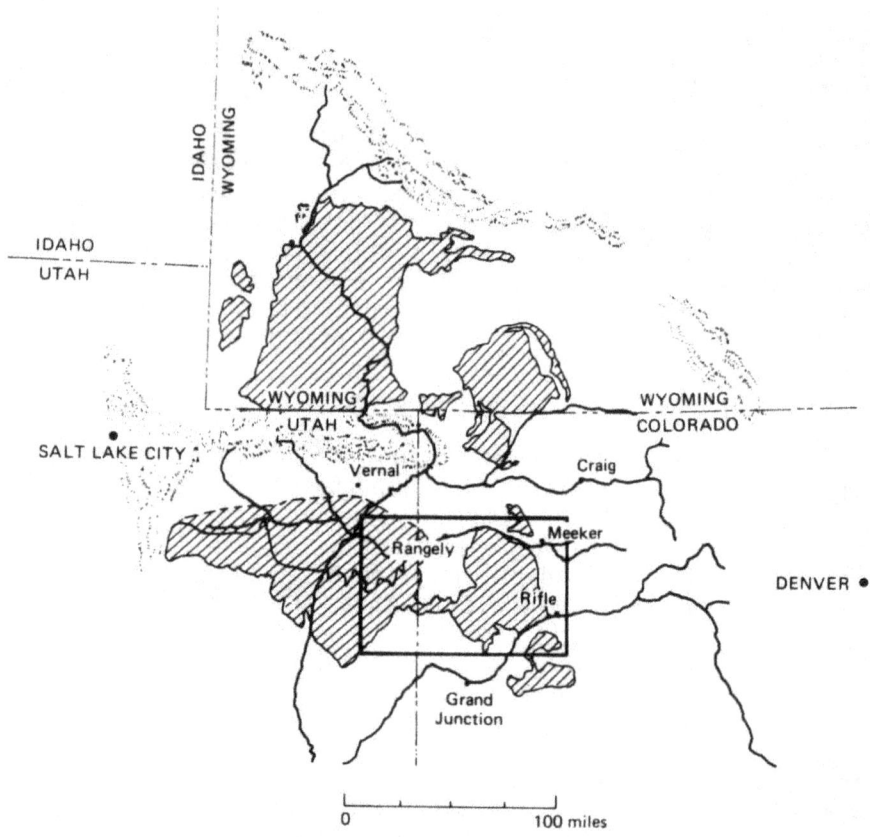

Fig. 1: Areas of oil shale (shaded) in portions of Colorado, Utah, and Wyoming. The boxed area delineates region shown in Figs. 2 to 5. Courtesy of University of California, Lawrence Livermore National Laboratory, and the Department of Energy under whose auspices the work was performed.

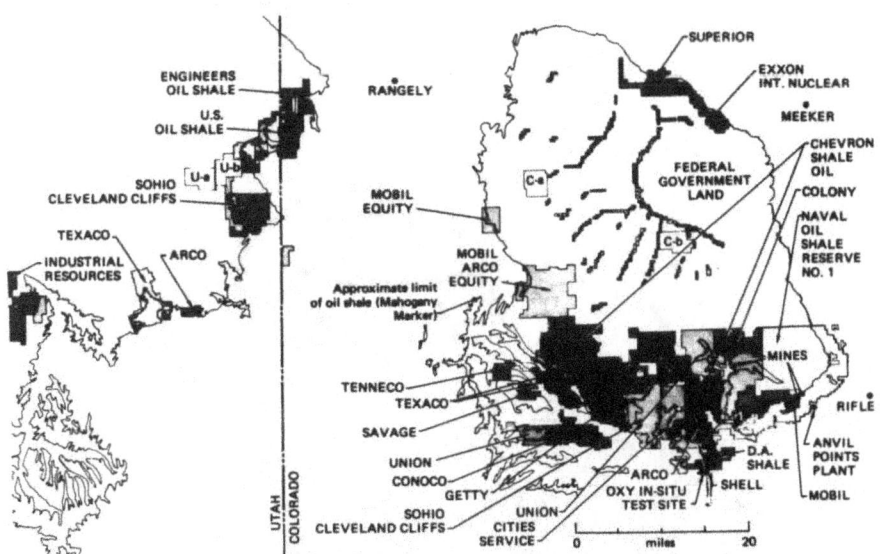

Fig. 2: Oil shale of economic interest and land ownership in Piceance Creek Basin in Colorado and a portion of Utah. Base map abstracted and reduced from original supplied by Cameron Engineers, Inc., Denver, Colorado. Courtesy of University of California, Lawrence Livermore National Laboratory, and the Department of Energy under whose auspices the work was performed.

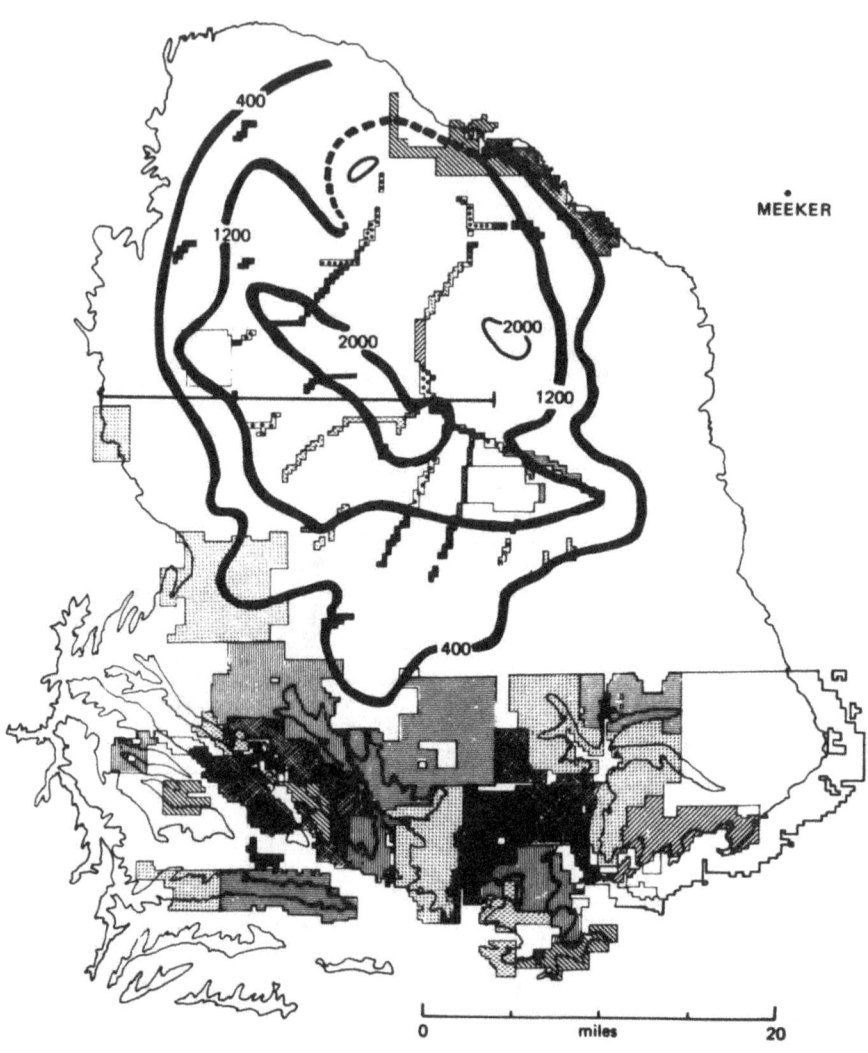

Fig. 3: Piceance Creek Basin: Contour lines show thickness
(ft) of the oil shale that contains 20 gal/ton of shale oil
(portion of base map of Fig. 2). Courtesy of University of
California, Lawrence Livermore National Laboratory, and the
Department of Energy under whose auspices the work was performed.

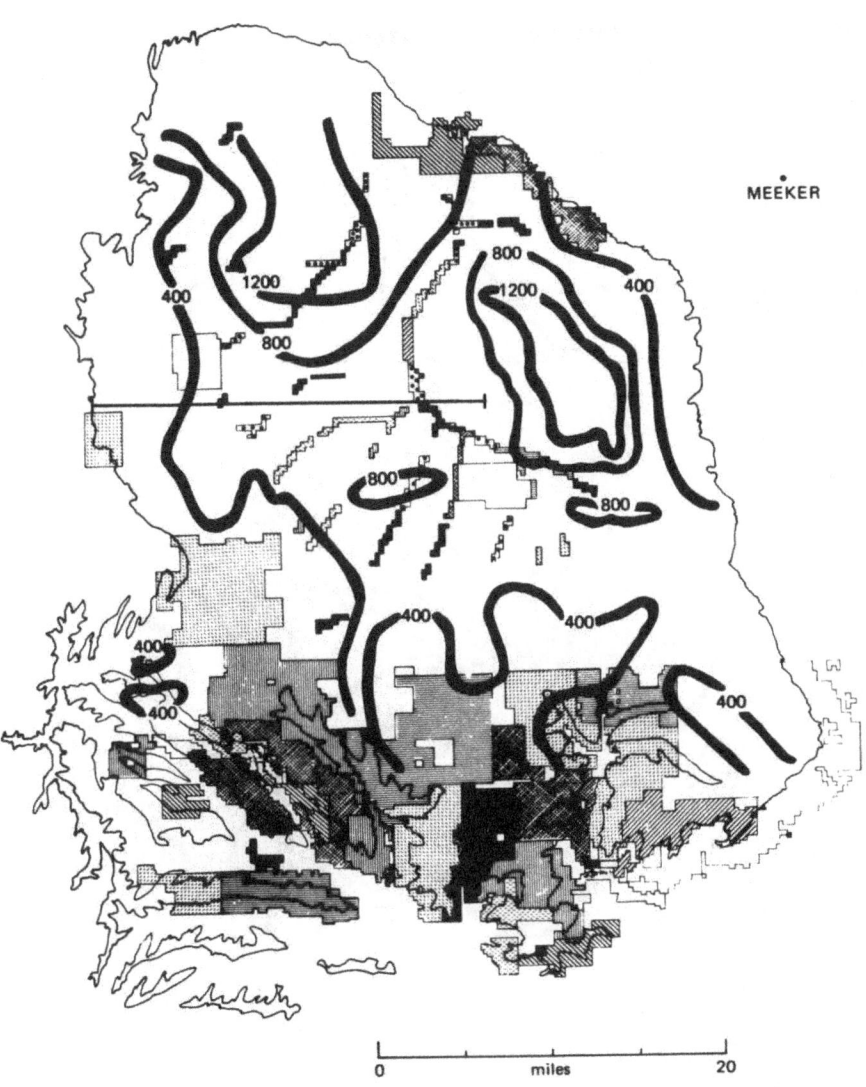

Fig. 4: Piceance Creek Basin: Contour lines show thickness (ft) of overburden covering oil shale of grade 20 gal/ton (portion of base map of Fig. 2). Courtesy of University of California, Lawrence Livermore National Laboratory, and the Department of Energy under whose auspices the work was performed.

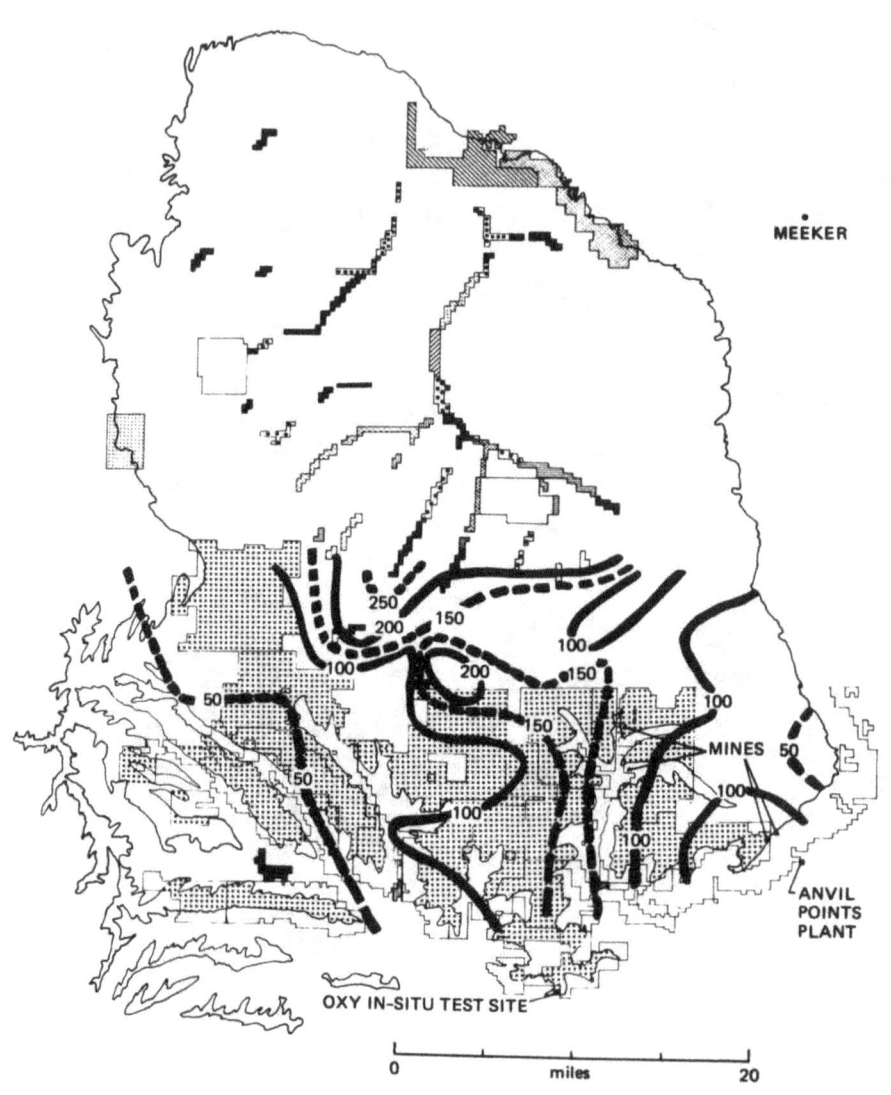

Fig. 5: Piceance Creek Basin: Contour lines show thickness
(ft) of oil shale containing 30 gal/ton of shale oil (Mahogany
Zone). Private lands containing oil shale are shaded (portion
of base map of Fig. 2). Courtesy of University of California,
Lawrence Livermore National Laboratory, and the Department of
Energy under whose auspices the work was performed.

on private lands contains approximately 50 billion barrels of shale oil. Oil shale above and below the Mahogany Zone of the privately owned property is too lean to be of interest. By contrast, in the central part of the basin, not only does the Mahogany Zone become thicker and richer but also the other zones, especially the deeper ones, contain economically important quantities of oil. Thicker sections of the Mahogany Zone are shown in Figure 5, on federal land north of the private land.

Historically, the southern part of the Piceance Basin was more accessible to exploration and potential development than the less well-known center of the basin, and it was contemplated that these beds were entirely suitable for mining and retorting. Thus, the southern portion was claimed by private parties, and the rich, deep part of the basin remains in government hands. The government land has been closed to claims since 1930 and, except for those plots acquired under the prototype leasing program of the Department of Interior, is not accessible to private parties. Under the prototype program, two plots of approximately 5000 acres each were leased in Colorado and two plots in Utah in 1974. Under the Mineral Leasing Act of 1920, no lease may exceed 5120 acres, and not more than one lease may be granted to any party.

The Green River formation also underlies large areas of Utah and Wyoming (Figure 1). None of this shale occurs in thickness and grade that averages more than 20 gal/ton and is more than 400 feet thick. High-grade layers in these regions are, for the most part, too thin or too deeply buried to be of serious economic interest in the foreseeable future. The exception is a strip in eastern Utah where the shale occurs close to the surface. Although it is generally thinner and lower in grade, this resource is similar to that in the southern part of the Piceance Basin and is suitable for room-and-pillar mining.

In summary, the bulk of the oil-shale resource exists in a small area in the north central Piceance Basin and is in government hands. A much smaller resource, in private hands, is in the form of thin high-grade layers, marginal by comparison, in the southern part of the Piceance Basin in Colorado and in eastern Utah (Figure 2).

Mining Technology

For an average 30 gal/ton shale, and using surface retorting, it will require that about 1.4 tons of shale be mined, transported, crushed, and retorted for each barrel of

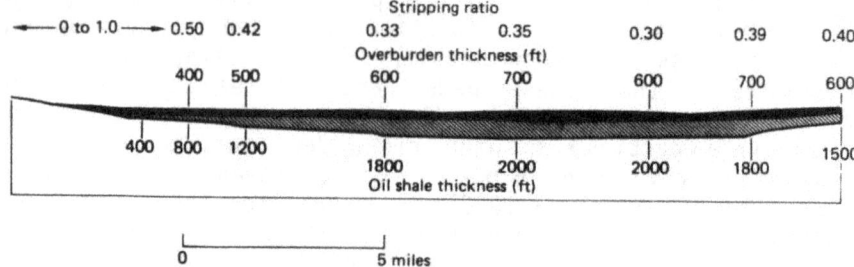

Fig. 6.   Section in the west central Piceance Creek Basin
          showing thickness (ft) of oil shale (20 gal/ton),
          overburden, and stripping ratio required for
          large-scale open-pit mining (see Fig. 3 or 4 for
          location).

shale oil produced. This means that a 50,000 barrel per day plant would have to mine 70,000 tons of shale per day. Either underground or surface mining can be used depending upon the thickness of the overburden. Hard-rock mining methods will have to be used which means room-and-pillar or block-caving underground mining and open-pit surface mining.

Underground mining methods are not as suitable as open-pit methods for recovery of a large fraction of this resource (9). No underground mining method is capable of mining very thick deposits at a low cost (10) and a high recovery factor; the best of these methods, such as block caving, even if applicable, would probably result in substantial surface disruption. Other considerations favoring open-pit mining are that fewer workers are required, lower accident rates are experienced, and required skills, such as power shovel and truck operators, are more easily obtained. The working environment is much better than for comparable production rate operations in underground mines. Skilled underground miners also take longer to train and work under more dangerous and unpleasant conditions. Recoveries of over 90% of the oil shale in place are possible by surface mining while recoveries of less than 60% are the best to be expected using underground mining. Environmental damage will also be much less for surface mining methods.

One of the locations most suitable for open-pit mining can be seen in a section through the central part of the Piceance Basin (Figure 6) which illustrates the thickness of the oil shale and overburden and gives the stripping ratio (9). The location of the section is shown in Figures 3 and 4. In selecting a starting site for the open-pit mine, an operator must choose between starting near the outcrop on the western end and starting farther east. The western site would permit early production without preproduction stripping, but the overburden ratio of one is higher and the pit would advance rapidly, thereby increasing haulage costs. Farther east, the stripping ratio is more favorable and pit advance would be slow, but preproduction overburden removal would be required.

Retorting Technology

The organic material in oil shale is not a liquid and it is not soluble to any significant extent in hydrocarbon solvents. The solid organic material is called kerogen and is contained in an impure marlestone having properties more similar to limestone than to a true shale. The rock must be heated to 900°F (450°C) or higher for the kerogen to undergo a thermal decomposition (pyrolysis) to yield liquid shale oil and hydrocarbon gas, leaving behind on the matrix a char composed mostly of carbon. A standard analysis called a Fischer Assay is used to measure the amount of oil (gal/ton or ℓ/Mg) produced from a given sample of oil shale. The actual amount of shale oil produced by various processes is then compared to that in this standard assay. The gas and char products are not counted, even though they may provide energy to run the process, or in the case of gas, may be collected and processed into a valuable product.

Some energy is absorbed in decomposing kerogen, in driving off the water contained in minerals, and in decomposing or reacting part of the carbonate minerals in oil shale. Additional energy is carried out of the retort by product oil and gas, by spent shale, and by combustion gas to the extent that these products leave the retort at temperatures higher than their entrance temperatures. The most efficient process is one in which both oil and gas production is maximized, decomposition of carbonates is minimized, and heat loss is minimized. If this can be done, char can be utilized efficiently to run the process, to produce hydrogen by reaction with water, and possibly to provide excess energy. The shale composition, heat sources, and sinks for an oil shale containing 24 gal/ton (100 liter/Mg) of oil are given in Table 1. The amount of kerogen is closely proportional to the grade of the richer or leaner oil shale, but the carbonate content is not directly dependent on either the kerogen content or grade.

All processes under consideration for commercial size module operations require that oil shale be divided or broken so heat can be applied to decompose the kerogen. In surface retorting processes, mined, crushed, and sized oil shale is heated in a retort by various methods to produce shale oil and other products. In modified-in-situ (MIS) processes, mining and blasting techniques are used to prepare underground chambers containing broken shale, which is then retorted in situ by using a combustion process. True in situ processes rely either on natural permeability or on permeability produced by using explosives to redistribute natural porosity. Heat is then provided by combustion underground or by introducing superheated steam or other hot gases which are heated outside the retort.

Surface Processes

There are three basic types of surface processes: (1) combustion retort (direct heated), (2) hot gas retort (indirect heated), and hot solid retort (indirect heated).

Table 1. Composition, Energy Sources, and Sinks for Shale
Containing 24 gal/ton of Oil (11)

Composition of Shale (wt%):

Carbonate	33
Kerogen	13.5
Water	1.0
Pyrite	1.5
Other minerals	51

Endothermic Reactions, "heat sinks" (kJ/kg shale)

Carbonate decomposition and/or reaction	325 to 650
Kerogen pyrolysis	46
Water release	23

Exothermic Oxidation Reactions, "heat sources" (kJ/kg shale)

Shale oil (C_5+)	3830
Iron sulfides	160
Char	960
Gas (C_4-)	390
Total exothermic	5340

Combustion or direct-heated retort. In a combustion
retort, crushed oil shale flows countercurrent to a stream of
air or other gases. At steady-state operation, oxygen in the
gases burns char and fuel-containing gases introduced into
the retort. The hot gases leaving the combustion zone heat
the oil shale and decompose the kerogen. Oil and hydrocarbon
gases are carried out of the retort where the oil is collected. The hydrocarbon gas is greatly diluted by the combustion products and nitrogen, and the outlet gas, therefore,
has a very low heating value, ranging from 80 to 120 Btu/scf.

Part of the retort exhaust gas is reintroduced into the
retort at controlled locations and in controlled amounts.
This recycled gas acts as additional fuel and as a control on
peak temperatures and their locations. This additional gas
improves the heat transfer to the shale and provides better
isolation of oxygen from the oil-producing zone. This increases the yield by reducing the burning of the oil.

A big advantage of the combustion retort is the thermal efficiency of the heat transfer between the solid and the gas. Both gas and solid leave the retort at reasonably low temperatures. Another possible advantage is the lower capital equipment and operating costs because additional equipment for heating the heat transfer medium and recovering energy is not required or is minimized. The combustion retort also has several disadvantages. The hydrocarbon gas must be used locally or burned in the retort because its low heating value prevents it from being transported economically to other areas. Only about half of the char is utilized as fuel in the process, and high combustion temperatures result in decomposition of some of the carbonate, thereby wasting energy. An additional problem is that the finest material from the crusher (5 to 15%) would interfere with the gas flow in the retort and, therefore, cannot be processed in this type of unit.

The major example of this direct combustion process is the Paraho (12) unit even though the Paraho retort can be designed and operated as an indirect heated process and several other types of units (Union, Petrosix, and Superior) can be operated in the direct-heated mode. Shale oil yields for direct combustion processes are reported to be as high as 90 to 95% of Fischer assay of the feed to the retort. The Paraho process uses gravity to move shale through the retort. Pilot units have been operated for extended times at a charge rate of 280 tons per day (t/d) of raw shale and well over 100,000 barrels of raw shale oil have been produced.

Hot gas or indirect-heated retort. The hot gas or indirect-heated retort uses a non-oxidizing gas heated external to the retort that passes through the retort countercurrent to the crushed shale to supply the heat necessary for the retorting operation. Yields of 95 to 100% can be obtained. The retort itself is similar to but less complicated than a similar type retort operated in the gas combustion mode. However, the process external to the retort is more complicated. The non-oxidizing gas is heated indirectly by burning recycle gas or the char on the shale leaving the retort. Shale leaves the retort at a high temperature near 900°F (450°C), and separate steps are required to recover this heat. However, temperature control in the retort is no problem and carbonate decomposition is avoided.

This process provides good control of the retort and high yields although at the cost of lower thermal efficiency and more complicated external equipment. Like the direct combustion process, it cannot handle significant quantities of fine material. Examples are the Union, Petrosix and

Superior processes. Union (13) uses a retort similar to the gas combustion retort but utilizes a "rock pump" to force shale upward through the retort. Superior (14) uses a horizontal moving grate to carry the shale through the retort. The Union retort (in the gas combustion mode) has operated at throughputs of 1200 t/d and the Superior retort at 250 t/d.

Hot solid retort. The hot solid processes utilize hot solids (spent shale or ceramic balls) heated outside of the retort as heat transfer media. This shot solid is mixed with raw shale in the retort and both the oil and gas are driven off and collected. The gas may be processed for sale or utilized on site. In the Tosco II (15) process, the gas is used to heat one-half inch diameter ceramic balls. These are mixed in the reactor with smaller sized shale and after leaving the reactor are separated and reused. In the Chevron (16) and Lurgi-Ruhrgas (17) processes the char on the spent shale is burned and the hot spent shale is used as the heat-transfer medium. In all of these processes large masses of hot solids must be processed and external heat exchangers used to recover the heat or the processes would not be thermally efficient. These processes have the advantages of high yields (90 to 100%) and the ability to process all the mined shale, including the fines rejected in the hot gas and gas combustion processes. Costs of crushing the feed to the smaller particle sizes are greater, however. The Tosco II process has been operated at rates near 1000 t/d and the Lurgi and Chevron processes at rates near 15 t/d.

Modified In Situ (MIS) Processes

The concept of preparing broken rock in underground chambers and retorting in situ is at least 50 years old. The advantage of this process is that only a part of the shale, 20 to 40%, is mined and brought to the surface in order to provide void space necessary for the proper operation of the retorting process. The mined shale is processed in surface retorts. The other 60 to 80% of the shale needs only to be broken and distributed uniformly to prepare it for in situ retorting. The retorting process is similar to the direct combustion process except that the shale remains stationary and the combustion and retorting fronts move downward through the shale. However, there are important differences.

The rock broken in situ cannot be carefully crushed and sized and the fines removed, as it is in surface processing. Larger fragments and a wider distribution of sizes must be processed. This requires a slowdown in retorting from 40 to 60 meters per day (m/d) to 1 to 3 m/d, or possibly less. Because of the grade and particle-sized distribution, the maximum recovery of shale oil expected is in the range of 55 to 70% for that part of the retort swept by the retorting front (18). The oil yield is further reduced if part of the retort is not swept by the retorting front. For example, if 60% of the retort is swept by the retorting front, the sweep efficiency is 60%, and the overall efficiency of recovery is the product of the sweep efficiency and the maximum recovery.

The gas flow through a retort must be uniform to obtain good sweep efficiency. It is essential that a uniform distribution of shale be produced to obtain a large yield of oil from the swept region of the retort.

There are other differences between MIS and surface retorting processes. The use of recycle gas to control temperature is not practical because the additional fuel slows the rate of retorting further. The entry point of the gas, or of the air, cannot be easily controlled, and the gas is a hazard underground becaue of its hydrogen sulfide and carbon monoxide content. Other gases such as steam are used to dilute the air to control peak temperatures.

Two MIS processes have been tested using large size in situ retorts, one by Cathedral Bluffs Oil Shale Company (19) and the other by Rio Blanco Oil Shale Company (20, 21). The major differences between the processes is in the methods of retort formation and rock fracturing.

Other Factors Impeding Development

Other factors also impede the development of oil shale. The oil shale of economic interest occurs in a relatively small area in northwestern Colorado with a small fraction being in eastern Utah. The area is semi-arid and sparsely populated. Development of shale oil requires water for production (mainly shale disposal and revegetation). In addition, people to operate an industry must be imported and provided with water, housing, roads, schools, and the like. Water rights probably sufficient for production on private lands have been reserved for many years, but this does not remove the controversy over water in Colorado.

Government regulation of many aspects of production and marketing is a fact of life. Much of this is appropriate but has not always been managed efficiently. The number of government agencies at the federal, state, and local levels that must be dealt with is overwhelming. The cost in time and money for obtaining not only the necessary permits but also a definition of requirements that must be met in the safety and environmental areas has become a major obstacle to develop-

ment. The potential for costly delay or even shutdown of a
facility during or after construction is a very real risk
that inhibits development. Some method must be found to
efficiently define the rules and conditions to be met so that
this resource can be efficiently developed with due regard to
safety, health, environmental effects, and cost of production.

Summary

     In summary, there is a strong case for continuing oil
shale development at the present time. Front end costs for
shale oil, per barrel per day for 20 years, compare favorably
with the front end costs of finding a new barrel of oil equivalent in the United States. Since direct operating costs
are estimated to be less than half the required price for a
barrel of shale oil, the total cost, including depreciation
and return on investment, will increase at a rate less than
the rate of inflation. Although risk is high because there
is no commercial size operation on which it can be evaluated,
this uncertainty is balanced by the well established
resources and the lack of finding costs.

     Unfortunately, the future of shale oil remains a ques-
tion mark. Over the past 20 years, we have been close observers of the roller coaster pattern with respect to oil shale
development and too many times have witnessed the on-again,
off-again situations caused by political and economic
crises. But the hard truth remains, we are still at the
mercy of the political situation in the Middle East and the
costs of finding and developing a barrel of new oil in the
U.S. is increasing at a much higher rate than inflation and
energy costs. We need to make a start on a commercial shale
oil industry to gain the know-how we need to provide for our
energy future. Shale oil is not the whole answer but it is
an important segment of the answer. Because of the long lead
times involved, we need to start our industry now. The willingness and ability of corporations acting in their short-
term interest will not necessarily fulfill the long-term
interests of the country. Once the flow of oil from the
Middle East is shut off, there will be very little time. The
strategic reserve of crude oil, even when filled, will last
only a few months. This may not be enough. It would seem
the time for a shale oil industry has come - will we realize
it?

References

1. Megill, R.E., "Problems in Estimating the Cost of
    Finding Oil and Gas," World Oil, 171-175 (May 1981).

2. Haun, J.D., Presidential Lecture, American Association of Petroleum Geologists, 1980.
3   McNally, R., Petroleum Engineer, 21-26 (January 1982).
4. Lewis, A.E., "Oil from Shale: the Potential, the Problems and a Plan for Development," Energy, 5(4):373-379 (1980).
5. Lewis, A.E., "Oil Shale: A Framework for Development," 13th Oil Shale Symposium Proceedings, Colorado School of Mines Press, Golden, CO, April 1980, 232-237.
6. Borg, I., Reconnaissance of the Oil Shale Resources of the Piceance Creek Basin, Colorado, from the Standpoint of In-Situ Retorting within a Nuclear Chimney. Lawrence Livermore Laboratory, UCRL-51329, Jan. 1973.
7. Lewis, A.E., Nuclear In-Situ Recovery of Oil from Oil Shale. Lawrence Livermore Laboratory, UCRL-51453, Sept. 1973.
8. U.S. Geological Survey, Denver, Colorado: PB 230-961, Average Oil Yield Tables for Oil Shale Sequences in Core from the Northern Part (TPS. 3S-4S) of the Piceance Creek Basin, Colorado, that averages 15, 20, 25, 30, 35, and 40 gallons per ton (1974); PB 230-962, Average Oil Yield Tables for Oil Shale Sequences in Core from the Central Part (TPS. 3S-4S) of the Piceance Creek Basin, Colorado, that average 15, 20, 25, 30, 35, and 40 gallons per ton (1974): PB 230-963, Average Oil Yield Tables for Oil Shale Sequences in Core from the Southern Part (TPS. 3S-4S) of the Piceance Creek Basin, Colorado, that average 15, 20, 25, 30, 35, and 40 gallons per ton (1974). Available from National Technical Information Service, 5285 Port Royal Road, Springfield, VA 22151.
9. Towse, D., Lawrence Livermore Laboratory, private communication based on data from J.K. Pitman and others (See also Refs. 2 and 3).
10. Hoskins, W.N., and others, A Technical and Economic Study of Candidate Underground Mining Systems for Deep, Thick Oil Shale Deposits. Cameron Engineers, Inc., Denver, CO, Final Report, Contract No. SO 241074, prepared for U.S. Bureau of Mines, October 1976.
11. Burnham, A.K., Lawrence Livermore National Laboratory, written communication, June 1982.
12. Jones, J. B., and R.N. Heistand, "Recent Paraho Operations." In 12th Oil Shale Symposium Proceedings, Colorado School of Mines Press, Golden, CO, August 1979.
13. Snyder, G.B., and J.R. Pownall, "Union Oil Company's Long Ridge Experimental Shale Oil Project." In 11th Oil Shale Symposium Proceedings, Colorado School of Mines Press, Golden, CO, November 1978.
14. Knight, J.H., and J.W. Fishback, "Superior's Circular Grate Oil Shale Retorting Process and Rundle Oil Shale Process Design." In 12th Oil Shale Symposium

Proceedings, Colorado School of Mines Press, Golden, CO, August 1979.

15. Baughman, G.L., Synthetic Fuels Data Handbook, 2nd Ed., Cameron Engineers Inc., Denver, CO (1978).

16. Tamm, P.W., C.A. Bertelsen, G.M. Handel, B.G. Spars, and P.H. Wallman, "The Chevron STB Oil Shale Retort," American Petroleum Institute, 46th Midyear Refining Meeting, May 1981.

17. Schmalfeld, I.P., "The Use of the Lurgi-Ruhrgas Process for the Distillation of Oil Shale." In Proceedings of the 8th Oil Shale Symposium, Colorado School of Mines Quarterly, 70 (3): 129-145 (1975).

18. Campbell, J.H., Modified In-Situ Retorting: Results from LLNL Pilot Retorting Experiments, Lawrence Livermore National Laboratory, Report UCRL-53168, 1981.

19. C-b Oil Shale Venture, Modifications to Detailed Development Plan (1 volume). Prepared by Ashland Oil Inc. and Occidental Oil Shale Inc., February 1977.

20. Rio Blanco Oil Shale Project Revised Detailed Development Plan, Tract C-a (3 volumes). Prepared by Gulf Oil Corporation and Standard Oil Company of Indiana, May 1977.

21. Rio Blanco Oil Shale Project Revised Detailed Development Plan, Tract C-a (1 volume). Prepared by Gulf Oil Corporation and Standard Oil Company of Indiana, June 1979.

Part 3

Use of Environmental Resources

David Abbey,
F. Lee Brown, Fred Roach

9. The Role of Water in Energy Development

Water resource problems were featured in many articles and broadcasts during the past year. Although these reports addressed a wide variety of topics, including the antiquated state of some urban water delivery systems, the potential for severe drought, the depletion of the Ogallala Aquifer in the High Plains grain belt, opposition to water storage or conveyance projects (such as the Peripheral Canal in California), and contamination of drinking water, they reflect two persistent themes: the prospect of water shortages and an impending crisis. As a prime example, consider a cover article in US News and World Report (1) entitled "Water: Will We Have Enough to Go Around."

> Suddenly hundreds of local water problems across the country are merging into one enormous national crisis. How people respond ... could have a profound impact on US economic growth and social structure in the years ahead.

Perhaps statements such as these reflect journalistic hyperbole. Nevertheless, they indicate an increased popular concern with water and the need for more technical information (2).

In this chapter we consider a subsidiary theme--the adequacy of water supplies to accommodate energy development in the semi-arid western US. This subject became prominent in the wake of the 1973 oil embargo and resurfaced with subsequent OPEC "price shocks," legislation to subsidize the synthetic fuels industry, and the current debate concerning coal leasing policy. Our purpose is not to review the technical literature. Rather, it is to address the layered perspectives on water and energy-related issues. The technical literature since 1973 reflects an evolution through these perspectives: from physical, through economic and institutional, to the emerging prescriptions for water resources management.

At its most general level, the water and energy issue is a physical problem. Water requirements for energy development may exceed the available supply. Throughout the seventies, numerous studies, often supported by the Federal government, identified either basins in which water shortages were likely to occur, or energy demand scenarios that were infeasible because of water supply constraints. For example, see (3,4).

If one views the subject of water and energy as a resource allocation or economic problem, more optimistic conclusions are compelling. First, the long-run price elasticity of energy demand is higher than commonly supposed in the seventies. Second, the economics of mine-mouth electricity generation are often unfavorable compared with coal shipments to the demand regions. Together, these two observations imply lower production rates than originally projected in the energy resource regions of the West. Third, analysts recognize that energy firms demand rather than require water; as water becomes more scarce, firms employ new water-conserving technologies. Fourth, alternative supply sources to unallocated surface water are available. More importantly, markets that contribute to the allocation of water resources have developed in the West. Fifth, the energy sector has a greater ability to pay for water or for conservation technologies than most other sectors have. In summary, these five considerations result in lower energy-related water demand in the semi-arid West.

In the second section, we review the economics of water use in the energy industry. We describe water demand and supply at the plant level and report results of more aggregrated, basin-level analyses.

Economic analysis of water use suggests generally favorable prospects for energy development in the West from a water-related perspective. But, each analysis also highlights the persistence of political and economic conflicts in water allocation and the need for institutional change. Institutional considerations, the focus of many current water- and energy-related studies, are the topic of the third section of this paper. In the fourth section, we make two tentative observations about remaining issues in the ever present conflict over water supplies in the West.

Water Use For Energy Production

The mix and quantity of factor inputs to production depend on the relative cost or availability of inputs. In most regions, water is inexpensive or even free. Not surprisingly, its use is quite intensive.

The budget for water use in the energy sector, compared for example to irrigated agriculture, is but a small fraction of total production costs. Capital and fuel costs dwarf water-use costs. Thus, the energy sector has an advantage in adapting to the new era of water scarcity. Energy can afford sharply higher payments to acquire water or capital investments (to conserve water) that have only slight or negligible effects on such fundamentals as selection of site, process, and output level.

In this section, we review water demand of and supply to the energy sector. To illustrate the interaction between energy development and water resources at the basin level, we present survey and modeling results.

Water Demand

There are four basic uses of water at energy conversion facilities: waste heat rejection or cooling, process use (as a boiler feed and a source of hydrogen for synthetic fuels), flue gas desulfurization, and solid waste disposal (in a slurry). Aside from the cost and availability of water, other factors influence water demand for these uses (5,6). These factors include production process characteristics, fuel quality (ash and sulfur content and heating value), degree of process water recirculation and reuse, the cost of water treatment technologies, residual discharge regulations, land disposal costs, and plant capacity factors. Process type and effluent discharge regulations are particularly important.

The production process determines the waste heat load and process water requirements, with considerable variation possible among processes. For example, the waste heat load at a coal-fired electric plant is one-third less than for an equivalent nuclear electric plant because of stack gas losses and slightly higher conversion efficiencies. A Lurgi coal gasifier can recover as much as 30% of the moisture in the raw coal feed whereas most second-generation gasifiers require a dry coal feed. Although the process type accounts for considerable variability in water demand, process selection normally is independent of water supply considerations.

Table 1. Water use for energy conversion[a] (A-ft/yr)

	Process	Cooling	Mining and Waste Disposal	Flue Gas Desulfuri- zation	Total
Coal Gasification (275 x 10^6 Mcf/d at 90% cf)					
Lurgi	550	5 050	1 350	800	7 750
Hygas	1 700	3 150	1 000	350	6 200
Coal Liquefaction (55 000 b/d at 90% cf)					
Synthoil	800	4 500	2 100	--	7 400
Shale Oil (55 000 b/d at 90% cf)					
Tosco	850	2 600	4 700	1 150	9 300
Paraho Direct	(350)	3 700	1 700	800	5 850
Electricity Generation (1 000 MWe)					
Coal (65% cf)	--	7 550	750	1 250	9 550
Nuclear (57% cf)	--	11 300	500[b]	--	11 800

[a]Water use for coal conversion calculated for a southwest site with sub-bituminous, high ash, low sulfur coal.

[b]Excludes water use for uranium mining and milling.

Legend: Mcf/d = thousand cubic feet per day
        b/d = barrels per day
       MWe = megawatt electric
        cf = capacity factor
      -- = not applicable

Source: Adapted from Probstein and Gold (10) and Abbey (6).

Design for zero-discharge or containment of liquid effluents at the plant site is standard practice throughout the West (7). In the Colorado Basin, this is due to effective prohibition of industrial salt loading. In other basins, it arises from anticipation that the Environmental Protection Agency will eventually promulgate zero-discharge regulations, or it arises from the desire to avoid National Pollutant Discharge Elimination System permit reviews and potential delays. Because the most efficient waste water treatment option, distillation, compares favorably with the cost of land disposal or solar evaporation, the zero-discharge constraint promotes the maximum degree of water recirculation, reuse, and treatment (7,8).

Consider again the four basic uses of water for energy conversion. Table 1 presents estimates of water use for seven energy conversion processes at standard-size mine-mouth facilities. These estimates assume extensive water treatment and reuse and are approximate upper bounds on water use for new energy conversion plants in the West. It is apparent that cooling water consumption is the principal target of water conservation in the energy sector. Dry or wet/dry cooling provides the demand-side response to water scarcity.

In physical terms, electricity generation provides the greatest potential for water conservation with dry cooling. Evaporation of cooling water accounts for 90% or more of total water use, excluding mine use. Table 2 shows alternative cooling system costs ($/kW and mills/kWh) and break-even water-use costs. Compared to bus bar electricity costs (about $1 000/kW and 30 mills/kWh), the incremental costs of 40% and even conventional 10% wet/dry cooling systems seem tolerable. However, the break-even costs are high compared to typical costs of water acquisition, treatment, and disposal for 100% wet cooling systems at most Western sites.

The Electric Power Research Institute and US Department of Energy are investigating advanced dry cooling concepts that use ammonia in a phase change process, enhanced heat transfer surfaces in the steam condenser, and deluge systems for partial wet operation (9). Advanced cooling technologies, which provide cost savings of about one-third compared to conventional systems, are nearing commercial availability. Firm estimates of break-even water costs for advanced dry or wet/dry cooling systems are not available, but they may fall below $300/acre-ft indicating great commercial potential for advanced dry cooling technologies in the electric utility section in the next decade.

Table 2. Cost of cooling alternatives at coal-electric plants for two sites in the western US ($1978)

	% Wet	$/kW	mills/kWh[a]	Break-Even Water Cost[b] ($/A-ft)
Farmington,	100	23	1.11	-----[d]
New Mexico	40	44	2.21	1 200
	10	57	2.86	1 570
	0[c]	48	4.07	8 770
Colstrip,	100	23	1.14	---
Montana	40	43	2.16	1 200
	10	57	2.68	1 260
	0[c]	47	4.02	9 560

[a]At 80% annual capacity factor; exclusive of water-use costs.

[b]Assumes evaporation of 0.45 gal/kWh with 100% wet cooling.

[c]High back pressure turbine used.

[d]Not applicable.

Source: Adapted from Hu, Pavlenco, and Englesson (11).

The potential for dry cooling at synthetic fuel plants is promising, even with conventional technologies because some waste heat loads occur at higher temperatures than the range typical of steam turbine condensers at electric plants. In fact, the water use estimates for synthetic fuel processes shown in Table 1 reflect extensive use of dry cooling even under the assumption that water supply is free. Cooling water consumption can be approximately halved from the estimates shown in Table 1 at incremental product costs of about 1% and break-even water supply costs of $80 to $1 300/acre-ft (10,12).

Water Supply

Although the hydrologic cycle is well known, for economic analysis it is convenient and sensible to consider water as a stock; it is sensible because that treatment is generally afforded by western state water law, by the doctrine of prior appropriation. One may identify four potential sources of supply to the energy sector: unallocated surface water, water in existing uses, groundwater, and waste water.

For many reasons, the quantity and price of water available in these supply categories are uncertain. First, there is a lack of data. In state water plans, for example, the data is fairly aggregate, omits price considerations, and provides superficial treatment of groundwater. Second, there is uncertainty concerning the definition of individual water rights, especially the consumption entitlement. Third, there are questions concerning the interpretation of existing law: for example, the degree of protection afforded adjacent or downstream water users in the case of a water transfer or application for a new groundwater withdrawal. Finally, there is the prospect of legislative change as a reaction to development. In any event, the steep break-even water-use costs for conventional, commercially available dry cooling technologies and the inelastic demand for process and other uses encourage energy firms to go to great lengths to acquire water.

Unallocated surface water is an increasingly rare phenomenon. The principal sources are existing or planned storage projects of the US Bureau of Reclamation, often at cost-based (inexpensive) prices of about $15/acre-ft. (This may be changing.) Although development of project water faces mounting obstacles from competing demands of the agricultural and municipal sectors and for instream uses (13), it is still common. Exxon, for example,

Table 3. Freshwater consumption in ten western states in 1975 (10^6 gal/d)

	Irrigation	Public Supplies	Rural Use (Domestic and Livestock)	Self-Supplied Industrial	Thermo-Electric Power
Arizona	5 400	200	66	210	41
Colorado	5 100	110	37	59	12
Idaho	4 700	34	27	160	2
Montana	2 700	49	55	12	nil
North Dakota	150	29	36	24	19
Nevada	1 500	52	14	71	22
New Mexico	1 400	83	56	85	33
South Dakota	180	14	100	6	3
Utah	2 200	130	14	51	8
Wyoming	2 000	46	26	34	24
Total	25 330	747	431	712	164
Per Cent of Total Use	92.5	2.7	1.6	2.6	0.6

gal/d = gallons per day.

Source: Adapted from Murray and Reeves (17).

recently signed a contract for up to 6 000 acre-ft/yr for its Colony oil-shale project.

In the future, water in existing uses will be the most important source of supply to accommodate development. Agriculture currently accounts for about 90% of water consumption in the West compared to less than 1% for the energy sector (see Table 3). In some basins, for high value crops like citrus, the value of water may approach $200/acre-ft but the marginal value in hay and alfalfa production (which predominates in the high-altitude, irrigated regions of the High Plains, Rocky Mountains, and Great Basin) is less than $10/acre-ft (14,15). For water-use reductions up to 40%, the marginal value in irrigation is still under $60/acre-ft.

North and South Dakota effectively prohibit transfer of water from irrigators to industrial users. In recent years, the Wyoming Board of Control denied or sharply reduced the quantity of water available as transfers to the energy sector (16). But there is evidence that such transfers are occurring throughout the West (Table 4) and that, even more frequently, the energy sector is purchasing and leasing irrigated land for future purposes.

The price of irrigation water varies widely depending on basin supply and demand. Generally one may say that farmers and ranchers value the market price of their water. In a well-publicized transaction, Intermountain Power Project purchased rights to 40 000 acre-ft in the Sevier Basin of Utah for $1 750/acre-ft (19). In a recent classified advertisement in the Wall Street Journal, a 3 000-acre cattle and sheep ranch in Rio Blanco County (the heart of Colorado oil-shale country) was listed. The ranch has "16 cfs early water," which exceeds 1 100 acre-ft/yr. If the entire value of the ranch were attributed to its water rights, that value would exceed $2 000/acre-ft, or $100-200/acre-ft on an annual, unit cost basis (20).

The stock of groundwater resources, compared to annual surface flows, is immense throughout basins of the West. The states' treatment of groundwater extraction varies considerably (21). Several states, including Montana and Nevada, restrict withdrawals to the rate of annual recharge. In Arizona many basins are closed to new appropriation. However, the energy sector is at considerable advantage because of its ability to pay. It can tap relatively deep (one thousand feet or more) or brackish aquifers, conduct hydrogeologic investigations, and thereby reduce or avoid interference with existing water users.

Table 4. Transfers of water rights to energy firms in the intermountain west

To	From	Quantity for Consumptive Use (A-ft/yr)
Colorado River Basin		
Utah Power and Light Company (Huntington and Emery Plants, Emery County, Utah)	Cottonwood Creek Consolidated Irrigated Company Ferron Creek Irrigation Company Emery County Water Conservation District (under contract from the Bureau of Reclamation)	5 000 7 000 6 000
San Diego Gas & Electric Company (Sundesert Plant, Blythe, California)	Metropolitan Water District Water rights obtained from purchase of 7 700 acres ranchland in Palo Verde Irrigation District	17 000 33 000
Nevada Power Company[a] (Reid Gardner, Moapa, Nevada)	Purchase of a ranch and leasing winter agricultural water	3 500
Nevada Power Company (2 000 MW, Las Vegas, Nevada)	Las Vegas and Clark County Sanitary Districts	43 764 (Sewage treatment plant effluent)
Arizona Electric Power Cooperative (350 MW, Benson, Arizona)	Purchase of 1 500 acres of farmland in Sulfur Springs Valley	7 000 (from wells)
Arizona Public Service Company (4 000 MW Nuclear plant, Wintersburg, Arizona)	City of Phoenix	64 000 (Sewage treatment plant effluent)
Great Basin		
Intermountain Power Project (3 000 MW, Lyndyll, Utah)	Shares in the Delta, Melville, Abraham, and Deseret Irrigation Companies and the Central Utah Canal Company	45 000
Arkansas Basin		
Public Service Company of Colorado[b] (1 000 MW, Las Animas, Colorado)	Los Animas Consolidated and Consolidated Extension Canal Companies	8 000 – 10 000
Platte River Power Authority (230 MW, Ft. Collins, Colorado)	City of Ft. Collins and Water Supply and Storage Company (a mutual ditch company)	4 200
Platte River Basin		
Missouri Basin Power Project (1 500 MW, Wheatlands, Wyoming)	Boughton Ditch, irrigated land inundated by by reservoir, and groundwater from Johnson Ranch	6 000
Panhandle Eastern Pipeline Company (coal gas plant, Douglas, Wyoming)	Douglas Reservoir Water Users Association (by financing repairs on a dam on La Prele Creek)	5 000

[a]In negotiation.

[b]Option agreement.

Source: Adapted from Abbey and Loose (18).

Potential sources of waste water include municipal sewage plants, uranium and oil-shale mines, and brackish return flows from irrigation. Compared to the other sources, potential waste water supplies are small, but such supplies match well the demands of the energy sector. Water-quality regulations often restrict the discharge of sewage or mine effluent. Each energy conversion facility can absorb flows up to 40 000 acre-ft/yr and can afford the investment in pipelines, reservoirs, and sediment treatment facilities.

At the plant level, the task of water acquisition seems tractable. However, it is also important to consider the aggregate water demands of the energy industry and the supply outlook at the basin level. The intense concentration of energy conversion plants in a few regions with relatively scarce water supplies may alter or qualify the favorable outlook for energy development.

Basin Analysis

We present two different approaches to basin or regional analysis: survey data of the current pattern of water use and results from an energy optimization model that incorporates water supply and demand. These approaches draw upon previous work examining the "water and energy" question.

Figures 1-3 present water-use data for electric generating plants projected to come on-line during the period 1980-1989 for selected river basins or states (22). Figure 1 shows that evaporative cooling continues to be the almost universal method of waste heat rejection. In fact, no commercial-scale sales of dry or wet/dry cooling systems to the utility industry are planned currently anywhere in the US. Figure 2 confirms that zero-discharge is standard practice in the West. New plants routinely operate cooling systems in the range of 10-25 cycles of concentration and reuse cooling tower blowdown for flue gas desulfurization or ash disposal (7). Figure 3 shows that, in the Colorado and Great Basins, utilities have turned to a variety of water-supply alternatives to surface water. By contrast, in the Upper Missouri Basin with its relatively abundant water supplies, surface water continues to be the favored source of supply. These survey data, which generally confirm findings drawn from consideration of water-use costs, indicate the response of the energy industry to water scarcity in different basins. Yet, one cannot blithely project these new patterns of water use for the future. Rapid growth of synthetic fuel markets, in

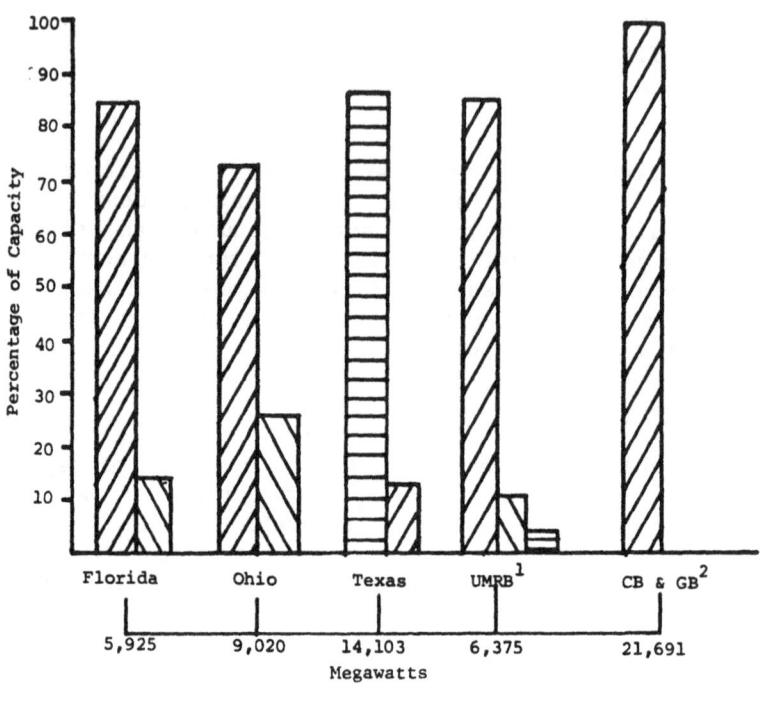

Figure 1.   Cooling systems at coal-fired and nuclear power
plants in selected states and river basins, 1980-1989.
Source: Abbey and Lucero (7).

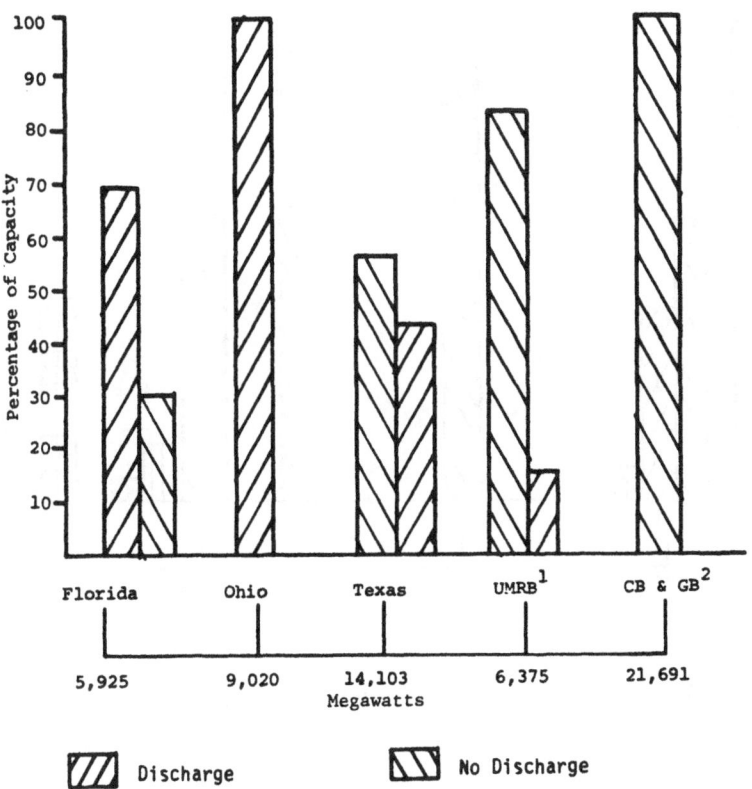

Figure 2. Discharge at coal-fired and nuclear power plants in selected states and river basins, 1980-1989. Source: Abbey and Lucero (7).

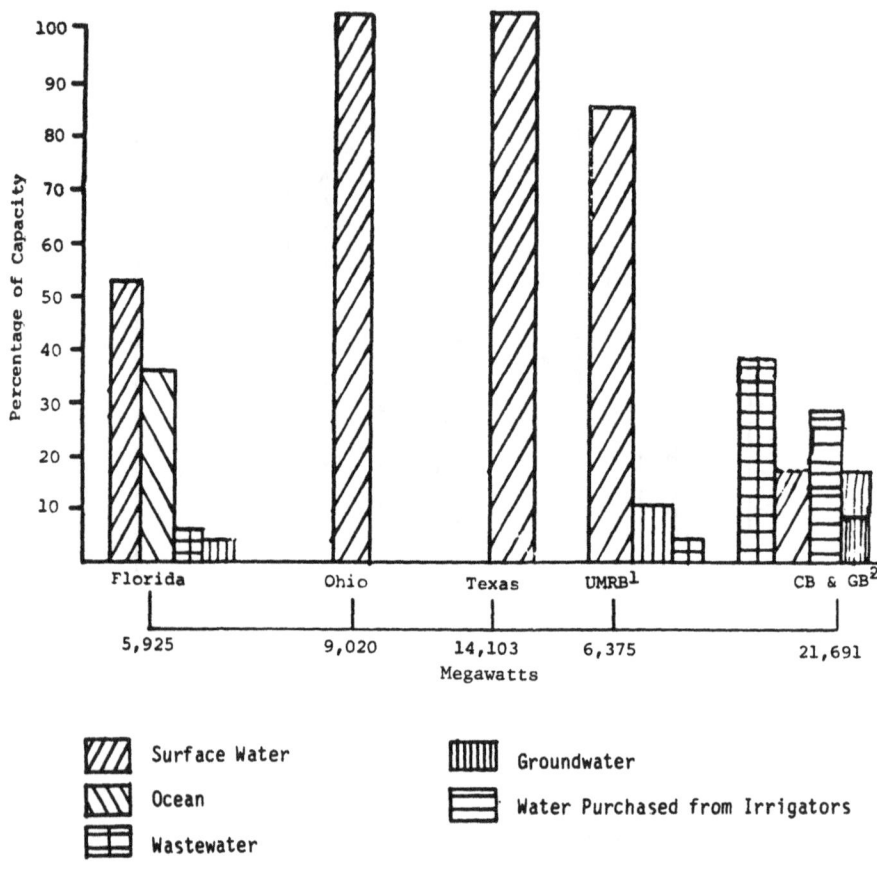

Figure 3. Water sources for coal-fired and nuclear power plants in selected states and river basins, 1980-1989. Source: Abbey and Lucero (7).

particular, might reverse the optimistic, short- to mid-term prospects.

To examine the relationships between the scale of energy development and basin suppliers, we used the Los Alamos Coal Use Modeling System (LACUMS). (The Appendix presents a more detailed description of the model and the scenarios.) The model includes a forecast of water use patterns in the energy sector to the year 1995 (18). The energy demand scenario for the LACUMS analysis included an effective annual growth rate in electricity consumption in the US of 4.5% from 1980-1995. Further, LACUMS included 2 quads (10^{15} Btu) of shale oil production, 1.333 quads of high-Btu gas from coal, and 0.667 quad liquids from coal, all from the West. We compared two water supply scenarios: a base case with water supply estimates as shown in Table A-1 of the Appendix and a more restrictive scenario with surface supplies available only in Idaho, western Montana, and North Dakota.

The difference between the value of the objective function (the minimum cost of energy production) in the two water-supply cases was only 0.6%. That figure applies to the total US, however, and is higher for the 10-state western region.

In both cases, about 33 000-MW coal-fired capacity and 1 300-MW nuclear capacity were sited in the 10 western states. Most of that new capacity was in the Southwest, reflecting demand growth in California and other Sunbelt states and the favorable economics of coal transportation by rail compared to mine-mouth generation and long-distance electricity transmission. Coal gas, and liquids facilities were sited in eastern Montana and western North Dakota, reflecting the abundance of low-cost, strippable coal and lignite reserves in the Northern Plains. In the more restrictive water supply scenario, about 3 000 MW of electric capacity shifted from eastern Nevada to the Utah portion of the Great Basin, and most of the coal gas plants in Montana shifted to North Dakota.

In both water supply scenarios, only 100% wet cooling was used at electric plants. Coal liquid facilities employed the maximum allowable fraction of dry cooling. Incremental dry cooling costs, however, were calculated for conventional technologies and may understate the potential for dry cooling with ammonia phase change loops.

Table 5 shows incremental water consumption for the two water supply scenarios for regional energy production in the

Table 5. Results of LACUMS analysis: water use for energy conversion in ten western states (10^3 A-ft/yr)

Coal Demand Region[a]	Type of Water	Base Case	Restricted Case[b]
29 (AZ-Salt)	Surface	24	---[c]
	Transfer	---	24
31 (AZ-NW)	Transfer	4	4
	Groundwater	38	38
32 (AZ-Yuma)	Transfer	131	132
39 (Idaho-Snake)	Surface	4	4
41/42 (MT-Lower	Surface	90	---
Yellowstone	Transfer	---	52
43 (NV-Elko)	Transfer	39	39
44 (NV-Truckee)	Transfer	72	67
45 (NV-Las Vegas)	Surface	44	---
47 (NM-Abq)	Surface	9	---
	Transfer	---	9
48 (NM-Lower Rio Grande)	Surface	18	---
	Transfer	---	18
58 (ND-Upper Missouri)	Surface	46	83
50 (Utah-Great Basin)	Surface	10	---
	Transfer	7	62
37 (Colorado-Yampa)	Surface	80	---
	Transfer	---	80
51/52 (Utah-East)	Surface	48	---
	Transfer	---	48
65 (Wyoming-Green)	Surface	16	---
	Transfer	---	16
Total	Surface	389	87
	Transfer	253	551
	Groundwater	38	38
Grandtotal		680	676

[a]Term(s) in parentheses indicate approximate geographic location of region.

[b]The surface water supply option is eliminated except in North and South Dakota and the Columbia River Basin regions of Montana and Idaho.

[c]None.

10 western states. Total incremental consumption is almost 700 000 acre-ft/yr. This is a relatively small amount compared to the approximately 25 million acre-ft/yr currently used for irrigation. For more disaggregated comparisons, the water demands of the energy sector still seem tolerable. As an extreme case, the model placed over 16 000 MW of new coal-electric capacity in Arizona (the figure is probably on the extreme high side) with concomitant demands for about 200 000 acre-ft/yr, representing less than 4% of water use in irrigation in that state.

One region where water scarcity presents a potential bottleneck to development is oil-shale country. The shale oil production scenario of 2 quad/yr is equivalent to about 900 000 bbl/day. With an 8 000 acre-ft/yr per 50 000 bbl/day Tosco II process facility, water consumption for shale oil production is about 150 000 acre-ft/yr. Because the richest oil-shale deposits are in Colorado, development places considerable pressure on the water resources of the Yampa and White River subbasins of the Colorado River (region 35 of Table A-1).

Process change might alleviate some problems. The Paraho direct process uses as little as 2 500 acre-ft/yr per 50 000 bbl/day plant and modified insitu processes may produce a surplus of mine water (23,24). On the other hand, shale oil demands are expected to grow rapidly around the turn of the century. While 2 quads a day may be optimistic for 1995, by year 2015 Exxon forecasts demand for 14 quads shale oil or about 7 million bbl/day (25). Unlike production of electricity or synthetic fuels from coal, tapping lesser grade deposits in other regions or transporting the oil shale are economically unattractive.

In summary, water acquisition by energy firms must overcome a variety of physical, economic, and institutional hurdles. Because of the variety of water supply and demand alternatives available to the energy sector, the physical and economic hurdles generally appear surmountable. This is particularly the case with coal-using sectors (electricity generation and production of synthetic gas and liquids) that have the additional advantage of siting flexibility. One may anticipate water-related constraints to shale oil production in the next century, but these are contingent upon uncertain technological developments and the persistence of restrictions on interbasin transfers. Only institutional considerations approach the status of a constraint to the energy sector. In a "crash" national drive for energy independence, such considerations are unlikely to affect the scale of development in the West as a whole, but rather direct development to or from certain basins or states.

Let us consider then the institutional framework governing water allocation with emphasis on those aspects related to energy development.

Institutional Considerations

Institutional Change

For our purposes, "institutions" refer to the entirety of laws, rules, administrative procedures, organizations, customs, habits, and other social forms that evolved to govern water allocation. The existing institutional machinery for western water was constructed gradually around the turn of the century for the principal purpose of appropriating virgin water and protecting established usufructuary rights. These institutional arrangements must now address the new tasks of reallocating water sources that are fully appropriated and of insuring the efficient use of increasingly scarce water supplies. These new tasks require institutional change at a minimum and to some extent the selective creation of entirely new institutions. Change can be observed throughout the region. Let us review some prominent examples, indicating the energy sector's reactions to and influences on the directions of change.

Instream Values. Laws requiring that water physically be taken from the streams to establish a beneficial use reflect the water development era in the West. With the advent of full appropriation, some states altered their statutory codes or judicial rules to confer legal status upon instream uses like fishing and canoeing, recreational or simply aesthetic appreciation. In addition, Federal legislation such as the Endangered Species Act has limited streamflow depletion. Recently, litigation between the Missouri Basin Power Project (MBPP) and the National Wildlife Federation (and other litigants) led to an injunction halting plant construction (26). MBPP settled out of court and agreed to curtail water use, modify reservoir operating procedures, cease further acquisition of irrigation rights, and establish a $7 million whooping crane habitat trust fund.

Increasingly, the states and the Federal government are comparing the value of traditional consumption of water with newly asserted instream uses. These comparisons imply additional risk and uncertainty for water and energy developers.

Water Markets. As long as water remained a commodity that could be newly appropriated by diverting a streamflow or sinking a well, there was little need for procedures allowing the buying and selling of water rights. Some states in protecting established rightholders during the development era even made water rights appurtenant to the land and legislatively prohibited their severance and transfer to other uses. But as new water demands arose in fully appropriated basins, transfers from existing users became a common source of water supply. This buying and selling of water rights has led to the development of rudimentary but recognizable water markets, with market specialists developing in some basins and states.

Many water transfers incur significant transaction costs. For example, in addition to the payments to irrigators reported above, the Intermountain Power Project (IPP) spent several million dollars for engineering studies and legal fees. The transaction costs associated with IPP's water acquisition average $75/acre-ft. Energy companies can afford the significant costs incurred in many water transfers. This leads to the clarification through case law of the terms governing transfers and the increased marketability of water rights.

Interstate Transfers. A major element underlying western water institutions during the development era has been state sovereignty over the water resources within its boundaries. When rivers such as the Colorado and Rio Grande flowed through or by several states, extended and expensive negotiations resulted in interstate compacts and judicial decisions dividing the expected flow of the river among the states. Thus, state sovereignty prevailed. Many state constitutions confer ownership of the waters within the state to the people of the state.

In recent years, this territorial supremacy over water has been assaulted. Two lawsuits, Colorado v. New Mexico (27) and El Paso v. Reynolds (28), if successful, would take water from a fully appropriated surface water basin and a closed groundwater basin, respectively, and move it for use in an adjoining state. According to a principal participant in the latter suit, a successful interstate transfer would undermine the foundation of the interstate compacts. Another example of the growing pressure for institutional change in this area is the persistent effort by Energy Transportation Systems, Incorporated (ETSI) to construct a slurry pipeline for shipping coal from Wyoming to Arkansas and beyond. After encountering difficulty in obtaining Wyoming water, ETSI recently reached a novel agreement with the government of South Dakota that may lead to the export of 50 000 acre-ft/yr of water from Lake Oahe and South Dakota sovereignty.

The capital cost of conveyance limits the frequency of such interstate ventures. However, the energy industry, with its considerable ability to pay for water, especially for coal slurry pipelines and oil-shale development, will be at the forefront of pressure to allow for interstate transfers. One consequence may be the evolution of stronger regional water management institutions.

Quantification of Reserved Rights. As long as unappropriated water remained during the water development era, both Indian and other Federal reserved water rights could remain unquantified without pressing too strongly on competing claimants for water. However, with the advent of full appropriation, existing appropriators increasingly comprehend the uncertainty that these paper rights pose for their own access to "wet water." The consequence has been increased interest in quantification (and therefore limitation) of reserved rights. Examples include the judicial decision in the United States v. New Mexico (29) and recent legislation (enacted and proposed) in Congress and various state legislatures. Some Indian leaders, recognizing the increased pressure, are concerned that litigation, legislation, and negotiation are inadequate and that the time has come for Indian tribes to exercise their rights.

Water Development Cost Sharing. A strong indicator of the transition from water development to water management as the central societal task is the impending change to a Federal-state, cost-sharing mechanism to finance future water development projects. Although the exact formula is still undetermined at this writing, bipartisan congressional bills garner even the support of legislators from the western states. During most of this century, when the reclamation ethic was dominant and an accepted societal objective was "to make the desert bloom," western politicians needed no "state cost sharing" to secure Federal funding. The subsidization of western water users, particularly in agriculture, was once the accepted political practice, but the future offers abundant alternative management techniques (30).

Fundamental Issues

During the transition period when new water management rules are being formulated and institutionalized, water users, particularly relatively new participants such as the energy industry, must recognize several related issues at the heart of water resource allocation. First, we consider objectives in water allocation. Is water (or, should it be) simply another commodity? Does water carry symbolic importance far exceeding its material value? Second, we review questions of conflicting claims to water ownership. Finally, we discuss the appropriate form of water management institutions. Alternative management forms range from pure laissez-faire market schemes to complete centralization of water allocation by state agencies or independent public corporations. There is no private or public concensus on these subjects at present, but developing attitudes will shape institutional conflicts and changes.

Societal Attitudes Towards Water. Some economists argue that water is like any other commodity. As it becomes increasingly scarce, it should be allowed to increase in market price and be allocated by market processes. Other students note that water is

> the object of a very complex structure of evalua-
> tions, rituals, superstitions, and attitudes. It
> has been the subject of sacred observances from
> very early times in human history.

The latter characterization (31) contributes to what is termed the "water is different syndrome" in which social attitudes require that water be treated differently than most other natural resources. A core element in this view is the indispensability of water to life itself. While there may be a high degree of substitutability in water uses such as the energy production technologies discussed earlier, it is inescapable that for basic life processes water must be present in biologically "fixed proportions." This core fact, combined with man's aquatic origin and agricultural heritage, easily accounts for a historically different set of social values being attached to water.

There is considerable evidence that this valuation structure survives today in the symbolic importance that Indian tribes attach to water rights (32) and the emotional intensity with which rural agricultural water users resist losing control of water. Water is valued not only for its historical importance and indispensability to life; in the semi-arid West it is also seen as the critical controlling element of economic destiny. Loss of control over water is seen as a forfeiture of future opportunity by those conditioned to periodic drought. Public attitudes in the less arid sections of the country do not appear as sensitive. But, faced with a future condition in which the demand for their native water exceeds their supply, the same latent valuations may manifest themselves. For example, in a recent public poll measuring attitudes on water, 70% of the respondents did not even wish to "consider selling any extra water" to Texas and Oklahoma (33).

Regardless of the depth and extent of the intangible social value structure that overlays the tangible substance, water, water remains increasingly scarce relative to the demands placed upon it. To the extent that water is important to the material well-being of society and that material well-being is socially important, water must be allowed, and even encouraged, to move to its highest valued economic use. To deny that movement is to forfeit the economic gains such movement makes possible. A corollary asserts that past practices of subsidizing water use must dissipate as a matter of public policy. Increasingly, water must be valued at its actual opportunity cost if it is to be managed wisely at all levels within the economy.

A minimal conclusion to the above discussion is that the evolving institutions for managing water (again contrasted with simply developing it) must take account of entrenched attitudes. If, on the one hand, these attitudes are viewed as anachronistic, then at a minimum, successful management institutions must incorporate a strategy for changing this element of the public attitude towards water. If, on the other hand, the view that "water is different" is accepted as supportable, or at least as given, then the evolving management structure must allow expression and some measure of control for proponents of this view.

Unresolved Ownership Problems. The most prominent problems of this type are the Winters doctrine claims of the various Indian tribes and other reserved rights advocated by the Federal government. Although many Indian leaders resist quantification as a diminution of their claim to water, the pressure for quantification is increasing. Even if the tribes successfully resist a fixed and final quantification of their rights, it seems likely that a minimum resolution of this question will require agreement on a formula for determining ownership. Current litigation utilizes the "practicably irrigable acreage" criterion promulgated in the Arizona v. California (34) decision of 1963. Although this criterion is an anachronism in light of modern economic conditions and although the inclusion of economic factors in the interpretation of the term practicably is resisted by Indian leaders, the criterion nevertheless provides a formula for determining the extent of Indian rights. Unless an alternative formula is proposed and agreed to by all interests, the pressure for elimination of the uncertain title to water created by the existence of reserved rights is likely to force quantification.

A second ownership question, not claiming public attention as forcefully as that of reserved rights, promises to play an even more fundamental role in the development of water management institutions. Is water public or private property? On its face, at least in the water law of most western states, this question is settled. Water has both a statutory and constitutional foundation in the law of most western states as belonging to the public with a usufructuary right granted to individuals to use the water for private purposes. For practical purposes, however, it is the latter title to water that dominates the actual allocation and control of water as well as the terms of compensation. Most state water administration institutions are confined to a regulatory authority to review private water transactions. Some structures, such as Texas groundwater law, do not even allow for this regulatory authority. Yet, there are signs of potential and growing conflict between these alternative institutional forms of ownership.

As long as new uses for water could be accommodated without retirement or threat to other uses, public sentiment tolerated a passive interpretation of public ownership. However, as full appropriation promotes reallocation of water, and as the economic value of water steadily increases, a more active assertion of public ownership and control may develop. The ETSI effort to obtain water for use in an interstate coal slurry line offers an example of this more active public role. In Wyoming, public action prevented what otherwise would have been a private transaction from occurring. South Dakota asserted an active public ownership because the negotiated agreement was with state government rather than with private parties, as a purely passive public ownership philosophy implies. Another example is a recent legislative proposal in Utah that would allow the State Engineer to consider the general economic benefit to the public in granting applications for Colorado River water. Such a criterion could reorder the queue of applicants currently temporally ordered by the date of application.

At this point, the debate over public versus private ownership is chiefly academic. However, increasing conflicts (35) are likely because it arises in large part from different societal attitudes towards water. The resolution of this fundamental ownership question will be central in structuring the form eventually assumed by water management institutions.

Water Management Institutions.   One could design a variety of management forms for the allocation and development of water if society were free of the existing institutional structure.   A key element in a management scheme is the locus for decisionmaking.   At one extreme, some philosophers argue for a pure laissez-faire arrangement in which decisions are made exclusively in voluntary bilateral agreements between individuals, with no individual having authority to bind others to an allocation without their explicit concurrence.   At the other extreme, one may idealize centralized decisionmaking.   This reflects an organic view of society in which achievement of collective social values is best accomplished through the socially binding decisions of a central unit.

Neither pure laissez-faire or complete centralization. is ever likely to be a practical scheme for managing water, and certainly society cannot design its institutional structure independent of the existing patterns and past history. Water development in the West exhibits elements resembling both laissez-faire and centralized decisionmaking.   Diver-. sion of "native water" as well as transfer of ownership and use have been largely a matter of individual initiative and action whereas "project water," particularly for irrigated agriculture, has required centralized funding decisions at the Federal level.   The pattern has been decentralized decisions for the water itself, and centralized decisionmaking for the allocation of capital to divert, store, transport, and apply the water.

In an era of broader water management functions, society must examine the suitability and synergism of these contrasting forms for modern tasks.   Moreover, the newly emerging water management institutions must be consistent with prevailing social attitudes towards the use and ownership of water.   Significant social conflict is likely as institutional changes emerge.   In certain states and basins, the institutional hurdle--from an energy perspective--may be severe(36).

Summary and Conclusions

As stated earlier, both the scientific and, to a lesser degree, the lay understanding of the relationship between water and energy in the West has passed through an evolutionary process.   In the crisis atmosphere engendered by the 1973 oil embargo, concern mounted over the inadequacy of the naturally occurring physical stocks and flows of western water to meet the large scale demands expected to arise from a burgeoning energy sector. This view yielded to an economic perspective in which reduced projections of energy development were coupled with an increased awareness of energy's considerable ability to pay for its water and the associated feasibility of large scale transfers of water rights. In this context, water has diminished as a regional constraint on energy development although local constraints still might be formidable.

The ability to pay conclusion, however, did not end the evolution in understanding. Although water is higher valued in energy uses and will "run uphill to money," societal concerns about the shifting ownership, control, and use of water have led to institutional conflicts that challenge the market directed movement of water. Moreover, the increasing value of water focuses attention on unresolved issues surrounding water in the West. Particular importance is attached to ownership questions, as embodied in Indian and Federal reserved rights, and to management questions such as state sovereignty in prohibiting interstate movement of water. Despite substantial evidence that the institutions governing water in the West are themselves evolving, significant problems remain and must be addressed.

We conclude this paper with two observations based on the preceding discussion. First, energy's use of water does not really present a unique set of problems. Instead, the key issue in western water affairs at this juncture is the changing nature of the western water institutions themselves. Although energy is a major actor in this evolving political environment, it cannot be treated in isolation from the broader context for water. Institutional dynamics influence, and are influenced by the energy sector's use of water.

Second, as the water institutions in the West are reshaped to perform water management functions in contrast to the more narrow water development tasks historically pursued, it is unclear to what degree active governmental intervention will be needed, particularly by the Federal government. Arguments exist for a substantial Federal role as trustee for Indian tribes, owner of reserved rights, arbiter of state disputes, and funder of development activities. However, counterarguments point to the need for decentralized basin or subbasin authority because the informational capacity to match societal purposes with the occurrence of the physical resource is greatest at lower governmental levels. The search for institutional solutions

Table A-1. Water supply estimates by coal demand region and water supply
category (10^3 A-ft/yr)

State/Region		Streamflows	Transfers	Groundwater
Arizona				
29	Phoenix (Salt)	240	1 091	---
30	Little Colorado	-----[a]	8	50
31	Colorado-Grand Canyon	-----	4	48
32	Colorado-Yuma	-----	876	180
33	Tucson	-----	123	103
Colorado				
34 38	Platte/Arkansas	-----	2 063	---
35	Green (Yampa/Whte)	98	113	17
36	Upper Colorado Mainstem	88	969	NA[b]
37	San Juan/Rio Grande	-----	812	NA
Idaho				
39	Central and Upper Snake	4 500	7 000	---
40	Lower Snale/Clarks Fork		No Constraint[c]	
Montana				
41	Columbia		No Constraint	
42	Upper Missouri	300	3 160	NA
61 62	Yellowstone	355	1 650	40
Nevada				
43	Great Basin (Elko)	-----	560	70
44	Reno (Truckee, Carson)	-----	275	335
45	Las Vegas (Colorado)	44	-----	---
New Mexico				
46	San Juan	80	80	2
47	Albuquerque (Rio Grande)	40	135	380
48	Pecos/Lower Rio Grande	70	1 500	400
North Dakota				
58	Upper Missouri	233	-----	40
59	Souris/Red		No Constraint	
South Dakota				
60	Upper Missouri	233	-----	40
Utah				
49	Weber/Jordan	-----	413	---
50	Virgin/Great Basin	10	1 902	85
51 52	Green/Colorado Mainstem	135	560	30
Wyoming				
53	Platte	-----	580	---
54	Powder/Tongue	217	151	40
63 64	Yellowstone	325	1 029	NA
65	Green	25	242	NA

[a]Nil.

[b]Not available.

[c]For several regions surface water supplies are taken as infinite. These
regions lack coal resources, are predominately rugged in terrain, and/or are
traversed by mighty streams.

Source: Adapted from Abbey and Loose (18).

to these counterforces is a prominent and difficult policy issue. It is very likely that the stresses will continue to grow before acceptable solutions are found.

Appendix

LACUMS is a partial equilibrium model of coal markets with particular emphasis on coal supply, electric utility capacity expansion, and environmental regulation of air quality and water quantity (37). LACUMS is solved through mathematical programming and is driven by exogenously specified energy demands and supply costs. The model is highly regionalized involving the division of the US into many coal producing regions, coal demand regions, and electricity consumption regions--in addition to the environmental regions associated with airsheds and river basins.

Water demand for electricity generation and coal liquefaction is treated as a three-step function with water conservation by partial dry cooling at higher costs (18). Water supply to the energy sector is described as a three-step function in 30 regions of 10 western states (Table A-1). The three steps represent potential supplies to the energy sector: unallocated surface water, transfered irrigation water, and groundwater. The acquisition cost of surface water is $20/acre-ft. Irrigation water costs $192.50/acre-ft (on an annual basis); groundwater, $211.75/acre-ft. The water quantity data were developed in an ad hoc manner by consideration of physical data in state water plans and Bureau of Reclamation planning documents, compacts allocating interstate stream flows, and state laws governing groundwater depletion and water transfers to industrial uses.

One must consider the results of the LACUMS analysis with some reservation. For example, it is a static analysis in which the demands of the municipal and non-energy sectors are fixed at current levels. There is a need to represent more steps in the water supply functions and to conduct sensitivity analyses of water-related input data. Nevertheless, it portrays the tradeoffs among the costs of energy transportation, water supply, and water conservation (dry cooling) and allows some comparison between the water demands of the energy sector and basin supply.

References and Notes

1. "Water: Will We Have Enough to Go Around," US News and World Report, June 29, 1981, pp. 34-36.

2. In fact, the theme of the 1982 American Water Resources Association Annual Meeting is "Water--Are We Running Out."

3. M. Harte and M. El-Gasseir, "Energy and Water," Science 199, 4329, 623-633 (February 10, 1978).

4. H. H. Hudson, "Ground Water Availability," Ground Water and Energy, Proc. of the US Department of Energy National Workshop, Albuquerque, New Mexico, January 29-31, 1980, pp. 37-47.

5. B. Bower, "The Economics of Industrial Water Utilization," Water Research, A. Kneese and S. Smith, Eds. (Johns Hopkins Press, Baltimore, 1966), pp. 143-173

6. D. Abbey, "Energy Production and Water Resources in the Colorado River Basin," Natural Resources J. 19, 2, 275-314 (April 1979).

7. D. Abbey and F. Lucero, "Water Related Planning and Design at Energy Firms," US Department of Energy report DOE/EV/10180-1 (November 1980).

8. J. Maulbetsch, "Water Quality Control Research at EPRI," presented at EPRI Workshop on Water Supply for Electric Energy, Palo Alto, California, March 19, 1980.

9. J. Bartz and J. Maulbetsch, "Are Dry-Cooled Power Plants a Feasible Alternative?" Mechanical Engineering, 103, , 34-41 (October 1981).

10. R. Probstein and H. Gold, Water in Synthetic Fuel Production, (MIT Press, Cambridge, 1978), pp. 263-277.

11. M. Hu, G. Pavlenco, and G. Englesson, "Water Consumption and Costs for Various Steam Electric Power Plant Cooling Systems," US EPA, EPA-600/7-78-157 (1978).

12. D. Abbey, "Water Use for Coal Gasification: How Much is Appropriate?" Los Alamos Scientific Laboratory, LA-8060-MS (October).

13. J. Thomas and D. Klarich, "Montana's Experience in Reserving Yellowstone River Water for Instream Beneficial Uses--Legal Framework," Water Resources Bulletin, 15, 1, 60-74 (Februarty 1979).

14. M. Gisser, R. Lansford, W. Gorman, B. Creel, and M. Evans, "Water Trade-Off Between Electric Energy and

Agriculture in the Four Corners Area," Water Resources Research J. 15, 3, 529-538 (June 1979).

15. R. Lansford, F. Roach, N. Gollehand, and B. Creel, "Agricultural Implications of Reduced Water Supplies in the Green and Upper Yellowstone River Basins," Vol I, Los Alamos National Laboratory, LA-8925-MS (March 1981).

16. F. Trelease, "The Changing Water Market for Energy Production," The Public Land and Resources Law Digest, 16, 2, 346-356 (1979).

17. C. Richard Murray and E. Bodette Reeves, "Estimated Water Use in the United States in 1975," Geological Survey Circular 765, USGS, Arlington, Virginia (1977).

18. D. Abbey and V. Loose, "Water Supply and Demand in an Energy Supply Model," US Department of Energy report DOE/EV/10180-2 (December 1980).

19. R. Clark, "Groundwater Conflicts and Barriers," Ground Water and Energy, Proc. of the US Department of Energy National Workshop, Albuquerque, New Mexico, January 29-31, 1980, pp. 49-64.

20. Further discussion of water markets and price data is offered in L. Brown, B. McDonald, J. Tyelling, and C. Dumars, "Water Reallocation, Market Proficiency and Conflicting Values," Bureau of Business and Economic Research, University of New Mexico, Albuquerque, New Mexico (June 1980).

21. G. Gould, "State Water Law in the West: Implications for Energy Development," Los Alamos Scientific Laboratory, LA-7588-MS (January 1979).

22. The survey included states in the East for comparative purposes.

23. T. Nevens, et al., "Predicted Costs of Environmental Controls for a Commercial Oil Shale Industry," Denver Research Institute, COO-5107-1, Denver (April 1979), pp. 227-318.

24. Mine water production might be attributed more properly to plant consumption. In fact, the mining of the shale zone, which is a saline aquifer, raises a variety of potential problems, including contamination of fresh water aquifers and sharp reductions in aquifer discharge and base flow of streams.

25. "The Role of Synthetic Fuels in the US Energy Future," available from Exxon Corp., Exxon Public Affairs Department, POB 2180, Houston, Texas.

26. W. Harrington, "The Endangered Species Act and the Search for Balance," Natural Resources J. 21, 1, 71-92 (January 1981).

27. Colorado v. New Mexico, No. 80, original in the supreme Court of the US, October term, 1977.

28. CN80-738-HB in the US District Court for the District of New Mexico. A similar case Sporhas v. Nebraska is on appeal to the US Supreme Court.

29. 438 US 696 (1978).

30. Consider, for example, the recent suggestion that hydroelectric and water storage projects be undertaken as joint investments by Federal, state, and private interests.

31. K. Boulding, "The Implications of Improved Water Allocation Policy," in Western Water Resources: Coming Problems and the Policy Alternatives, symposium sponsored by the Federal Reserve Bank of Kansas City (September 27-28, 1979), pp. 299-311.

32. For a mre complete discussion see P. Reno, Mother Earth, Father Sky, and Economic Development (University of New Mexico Press, Albuquerque, 1981), pp. 46-64.

33. Arkansas Public Awareness Survey on Water Resources, conducted by the Center for Urban and Governmental Affairs, University of Arkansas, Little Rock, Arkansas, July 1981.

34. 373 US 546 (1963).

35. In the El Paso v. New Mexico court case cited earlier, New Mexico's brief argues that water is publicly owned in New Mexico in contrast to the private doctrine that underlies Texas water law.

36. See, for example, two recently completed studies specifically addressing institutional considerations in water acquisition and management: G. Weatherford, "Western Water Institutions In A Changing Environment," Vols. I and II, project completion report by the John Muir Institute, Inc., for the National Science

Foundation, December 15, 1980; and G. Weatherford, "Acquiring Water for Energy: Institutional Aspects," prepared by the Center for Natural Resource Studies of the John Muir Institute, Inc., for the US Department of Energy, Fall 1981.

37. F. Wolak, R. Bivins, C. Kolstad, and M. Stein, "Documentation of the Los Alamos Coal and Utility Modeling System" Los Alamos National Laboratory LA-8863-MS (May 1981).

Charles D. Kolstad,
William D. Schulze, Michael D. Williams

10. Clean Air and Energy: From Conflict to Reconciliation

Congress is currently reviewing the Clean Air Act of 1967 and its amendments of 1970 and 1977, the body of legislation protecting the nation's air quality. The act is vigorously attacked by some and strongly supported by others. Some criticism is leveled at governmental regulation in general; the contention is that regulation is becoming an excessive burden to society. Other criticism of the act maintains that some of our economic ills are due to an overly strict Clean Air Program. For example, some contend that energy development in the western US is being severely and unnecessarily hampered by provisions of the act that prevent the significant deterioration of air quality.

Opponents in the debate have valid points. High air quality is important in the western US; it should be protected. On the other hand, energy sustains the US economy; our nation is strained by the rapid price rises for energy over the past decade. Because of our reliance on imports, our economy is vulnerable to the whims of foreign oil suppliers. If the western US can play a major role in ameliorating national energy difficulties, overly stringent air pollution regulations should not be permitted to impair energy development in the West.

The purpose of this chapter is to reconcile the apparent conflict between air quality protection and energy development in the West. After reviewing the nature of the western pollution problem, we discuss in turn the industrial costs of air quality protection and the monetary value of the damage to the environment by air pollution. Reconciling these costs and damages in a study of the Four Corners Region of the US, we conclude that high levels of energy development are consistent with the Clean Air Act and that industrial costs of current regulations are significant but modest relative to the value of the energy produced. If we consider a balance of costs and damages, current regulation appears to undercontrol sulfur dioxide (SO_2) emissions.

The Nature of
the Western Pollution Problem

Environmental Impacts of Degraded Air Quality

There are basically three ways that air contaminants affect the environment. The first occurs with toxic concentrations damaging the health of plants or animals. The second relates to effects of altered light transmission through the atmosphere. The third concerns deposition of materials onto sensitive surfaces or deposition and subsequent transport into soils or aquatic ecosystems.

Toxic concentrations may result directly from emissions of pollutants such as carbon monoxide or indirectly as in the case of oxidants. Pollutants may be primarily local in character or they may occur over very large areas. Furthermore, the concentrations may display a steeply sloping frequency distribution curve or a very flat one. Energy sources tend to have different impacts associated with tall stacks as opposed to fugitive sources or secondary development.

Concentrations of the oxides of sulfur and nitrogen and fine particulate matter associated with tall stacks are apt to be very sensitive to terrain and produce relatively high ratios of maximum short-term average concentration to annual average concentrations. In the West the highest concentrations are likely to occur under stable conditions on terrain near effective stack height, that is, 300 to 500 meters above the stack base. The presence of intervening high terrain can greatly reduce concentrations. Relatively high concentrations may occur at distances of 30 to 50 km from the source if the terrain and meteorological conditions are appropriate. Frequently, areas receiving high concentrations are relatively unpopulated although some of these areas are in national parks where people may demand better air quality.

Fugitive or secondary sources tend to produce high, but relatively local, concentrations of particulate matter and possibly carbon monoxide and sulfur oxides. In these instances the concentrations display much lower ratios of the maximum short-term concentration to the annual average. These concentrations occur at the source height.

In most of the West, ambient concentrations of most pollutants are low compared to air quality concentrations permitted by standards designed to protect health and plant life. Exceptions are relatively widespread occurrences of moderate levels of ozone, occasional high levels of particulate matter, and local high concentrations of carbon monoxide associated with automobile traffic or fireplaces and woodburning stoves. Communities in valleys frequently experience periods of low ventilation that rival the infamous meteorological conditions of Los Angeles.

High levels of particulate matter may be associated with wind-blown dust, mining operations, dirt roads, or woodburning stoves and fireplaces. In most areas short-term air quality standards for particulate matter are exceeded occasionally.

Deposition of the oxides of sulfur and nitrogen and their transformation products may change the acidity of high mountain soils and waters or damage plant tissue. Local areas near sources may receive high levels of deposited materials, but major mountain ranges are also major receptors. The mountain ranges produce most of the run-off and thus gather pollutants through wet deposition. Furthermore, rugged terrain tends to enhance dry deposition, and the mountains have more vulnerable soils and vegetation.

Altered light transmission through the atmosphere can affect traffic safety, plant growth, or aesthetics. In the West effects on aesthetics are the best documented (1). The Southwest generally has the best visibility in the contiguous states although good visibility also occurs throughout the mountain West and in the western portions of the plains states. The southwest and mountain states also devote large areas to parks and wilderness.

There are principally two types of visibility effects: regional haze, in which distant features appear indistinct but the contaminants are not readily identified with any single source or complex of sources; and plume blight, a gray or brown smear across the landscape, with an identifiable, single source.

In the instance of regional haze the most important contaminants appear to be sulfates (1,2). On occasion, nitrates, wind-blown dust, or carbon-based aerosols may be important. In regional haze, aerosol concentrations usually vary only slowly within the mixed layer. Furthermore, principal contributors to regional haze may be hundreds of kilometers from the point of observation. For example, with a source emitting an annual average of one million metric tons of SO_2 in California, the sulfate concentrations ($\mu g/m^3$) would be 1.6, 0.52, 0.45, 0.20, and 0.17 in California, Arizona, Utah, Idaho, Montana, and Wyoming based on a model developed by Brookhaven National Laboratory (3).

Nitrates have the potential to play an increased role in visibility in the future. First, oxides of nitrogen (NO_x) emissions are expected to increase more rapidly than SO_2 or fine particulate emissions. Second, gaseous nitrates have very little effect on visibility but particulate nitrates can have significant effects. At low concentrations almost all nitrates are in the gaseous form, but at higher concentrations particulate nitrates form an increasing fraction of total nitrates. For this reason, there is a potential threshold for effects of nitrates on visibility.

Concentrations associated with visibility impacts during regional haze can be quite low. For example, the addition of about 0.3 $\mu g/m^3$ of sulfate could reduce visual range by about 30 kilometers (4) and studies (5) show that such change is significantly valued by residents. In comparison, the annual average standard for particulate matter is 60 $\mu g/m^3$ while the ambient standards for the precursor SO_2 is 80 $\mu g/m^3$. Thus, sources producing little impact in terms of toxic concentrations may yield large impacts on visibility.

Plume blight is a more local phenomena. During stable atmospheric conditions, pollutants confined in a shallow layer of the atmosphere, may be visible at distances on the order of 100 kilometers downwind. Either upwind high terrain or changes in stability can eliminate visual impacts. Primary particulate matter and nitrogen dioxide tend to be the most important pollutants in plume blight. Low level and fugitive sources tend to have very local impacts while emissions from tall stacks produce impacts at greater distances. Because the bulk of the pollutants is well above the ground, ground level concentrations in plume blight are usually very low and there is no relation between plume blight and ambient concentrations.

One other major group of sources influences air quality in the western regions. The copper smelters in

Arizona, New Mexico, Utah, and Nevada have been associated with visibility impacts in the mountain West (1, 2, 4, 6). The smelters, large emitters of SO_2, have lower costs of control per unit of SO_2 removed than do power plants (7). Smelter emissions were approximately 2 million tons per year in the period from 1971 to 1974, but they have declined to about one million tons per year of SO_2 emissions.

Energy and the West

Air quality, a highly protected value in the western US, might be less important but for the extent of potential energy development planned for parts of the West. A very small area of the West covering parts of Colorado, Utah, and Wyoming contains essentially all of the high grade oil-shale resources of the US. Whereas forecasts vary, the Department of Energy anticipates shale oil production will meet the level of 200 000 to 700 000 barrels of oil per day (3). Other forecasts run into millions of barrels per day by the end of the century. Compare this to 1980 total US oil consumption of about 15 million barrels of oil per day (3). The oil-shale industry has the potential to grow very large and to place stress on air and other resources of the region.

Coal-based energy production has a large potential in the West. The western US contains over half of the nation's coal and over 90% of the nation's low-sulfur strippable coal (8). Western coal production is continuing to rise rapidly, particularly in the states of Wyoming and Montana. Aside from being shipped to the Midwest and East for combustion in power plants, coal can be converted at the mine to either electricity or synthetic gaseous or to liquid fuel for subsequent transport to population centers. The potential exists for large-scale coal conversion in the mountain west, particularly to serve the population centers of the Southwest and Pacific coast.

There are primarily three major new energy sources in the West. These are coal-fired power plants, coal syngas or synliquids, and oil-shale retorts with support facilities including mines, waste piles, and transportation networks for each. Although high concentrations of particulate matter may be found in the neighborhood of a mine, the mines have relatively local effects.

Coal-fired power plants emit primarily the oxides of sulfur and nitrogen and fine particulate matter. Control devices exist for both particulates and sulfur oxides, but oxides of nitrogen are controlled only by combustion techniques. In Japan, control devices for the oxides of nitrogen are being used for oil- and gas-fired installations, but techniques for coal-fired plants are less well developed. Consequently, a new, well-controlled plant will emit somewhat more oxides of nitrogen than sulfur dioxide.

Coal-fired power plants are relatively flexible in siting because coal and the needed water may be moved large distances to the plant site. In this context water is more expensive to transport than coal.

Coal synfuel plants also emit particulate matter, NO_x, and SO_2, but they tend to have relatively higher emission of hydrocarbons. For this reason they are likely to have more impact on ozone formation. However, for the same amount of coal processed, synfuel plants tend to have much lower emissions than coal-fired power plants. Emissions of NO_x from synfuel plants are much less likely to produce visible plumes because of the lower emission rates. Synfuel plants are also relatively flexible in their siting.

Oil-shale facilities tend to have emissions comparable to synfuel plants for the same energy input. However, oil-shale facilities have much less siting flexibility because of the large amounts of material they process.

Regulatory Background

Clean Air Act amendments of 1977 use a two-tiered approach for the protection of air quality in areas that are currently cleaner than ambient standards (9). If levels in excess of ambient standards occur in a region, non-attainment is reached, and provisions are activated to achieve the standards.

Of more concern to new sources in the West, are the provisions applying to areas with better air quality than the standards. In such areas the so-called Prevention of Significant Deterioration (PSD) provisions apply. Under the PSD provisions all new sources within major source categories may add specified amounts, PSD increments, to the concentrations of pollutants measured or estimated to exist in a baseline period. The increments are defined on the same basis as are the ambient standards, and thus for SO_2 there are 3-hour, 24-hour, and annual average increments. There are increments established only for total suspended particulate matter and sulfur oxides.

Increments are established for three different classes of area. Class I refers to national parks, national wilderness areas, and other areas for which the highest possible air quality is desired. Increments are only an intermediate step in determining construction of a new source. If the new source, in combination with other post-baseline sources, is expected to exceed permitted increments in some Class I area, the source may not be built unless the applicant demonstrates to the Federal land manager that there will be no adverse effect on air quality related values. If the source is projected to comply with the increments, the source may be constructed unless the Federal land manager demonstrates that there will be adverse effects on air quality related values.

Class II areas apply to most of the country where moderate industrial growth is desired. Currently this represents the entire country except for Class I areas. Class II increments must be met if a source is to be constructed. Class III increments apply to areas where large amounts of growth are desired and where air quality is a secondary consideration.

The PSD provision also requires that best available control technology (BACT) be used as determined on a case-by-case basis. However, in some jurisdictions BACT is assumed to be New Source Performance Standards plus whatever is required to meet increments.

The second tier of the approach restricts emissions independent of the location of the source. BACT is one such provision; New Source Performance Standards, which define the effectiveness of the control equipment that must be used, is another.

No specific provision sets a limit on SO_2 emissions although the increments and the BACT provisions reduce the rates of growth of this pollutant. The effects of the ambient provisions are further enhanced through requirements limiting the height of the stack used in estimating air quality.

One area of ambiguity in the regulatory process is fugitive dust. In the past fugitive dust was not considered in increment consumption determinations; however, after the policy changed, all emissions that can be reliably estimated are to be considered. This is potentially very important because the emissions associated with a dirt road are usually enough to exceed the Class II increments.

Obviously, consideration of low level sources could pose a major barrier to development of all kinds.

Another difficulty in the regulation of particulates is non-attainment. Under the fugitive dust policy of the Environmental Protection Agency (EPA), occasional high concentrations of particulate matter are ignored unless there are industrial facilities or urban areas nearby. Thus, the construction of a new source is, in theory, sufficient to create a non-attainment area from an attainment area. In practice these considerations are usually ignored.

The Cost of Pollution Control

In this section we review the costs associated with air pollution control. We first consider the general options open to industry to reduce air pollution impacts. Second, we present results of an actual analysis for part of the West in which we show the total costs of reducing sulfur pollution.

Air Pollution Control Options

This section illustrates the variety of options open to industry to reduce pollution impacts. We wish to emphasize that pollution control is a continuous process in the sense that emission levels can almost always be reduced at some additional cost.

Most commonly, pollution control is thought of in terms of adding a "cleaner," such as a scrubber for a power plant, to the emission stream of a facility. However, other options are open. In fact, there appear to be four basic options open to industry to reduce air pollution: add-on pollution control, location shift to reduce impacts, process change, or reduced operations. If pressures on air resources in one area are particularly strong, then industry can move to another otherwise second-best location that permits more pollution. Industry can choose to change production processes to those inherently cleaner. A fourth "control" measure is exercised by consumers: as products become relatively more expensive, due to pollution control, consumers may shift their demand from such products towards less costly ones (that is, inherently cleaner to produce).

We now concentrate on these points in the context of electricity production, production of "synthetic" fuel from coal (liquid and gas), and oil-shale processing.

Add-On Pollution Control. One approach to control the deleterious effects of pollution is to add equipment for removing pollutants from the waste streams of a facility. The SO_2 scrubber, for example, is typically affixed to coal-fired electricity generation plants. To reduce emissions, a variety of equipment can be appended to production facilities currently in common use to reduce emissions. However, this may not be cheap. Figure 1 shows the approximate effects on electricity-generating costs from various levels of SO_2 control for coal-fired power plants. Pollution control costs can be quite moderate if control is relatively loose, or costs can be large if control is tight. Synthetic fuels facilities face similarly higher costs with more efficient treatment of waste gases.

Location Shift. Air pollution legislation in the US recognizes locational differences. In some areas, such as national parks, very little incremental pollution is tolerated; other areas can absorb large increases in pollution without violating ambient air quality regulations. When we read that air quality regulations have prevented construction of a power plant at a particular location, we often neglect to appreciate that at modest additional cost, the power plant site can usually be changed (and often is) to avoid violating air quality regulations (11).

In power production, additional costs from location shifts result from increased transmission distance for power and, when coal is involved, from increased shipment distance for that coal. Moving synthetic fuels facilities similarly results in increased costs associated with moving feedstock, whether coal or shale and, to a lesser extent, moving product an additional distance. The actual micro-scale location choice, of course, involves other factors such as water availability; nevertheless, these costs associated with additional transport are the major contributors to cost increases. Figure 2 shows some very approximate costs that might be incurred in moving an energy facility. For instance, a power plant could incur additional costs of as much as 1/3¢/kWh (10%) by being forced to relocate 100 miles from a prime location (12). On the other hand, this move could greatly reduce some of the air quality problems. Synthetic fuels plants similarly could be expected to incur additional costs from a location change.

Process Change. Production processes naturally evolve to conserve scarce resources and utilize abundant resources. This suggests that the most cost effective way to control pollution is probably not just to add control

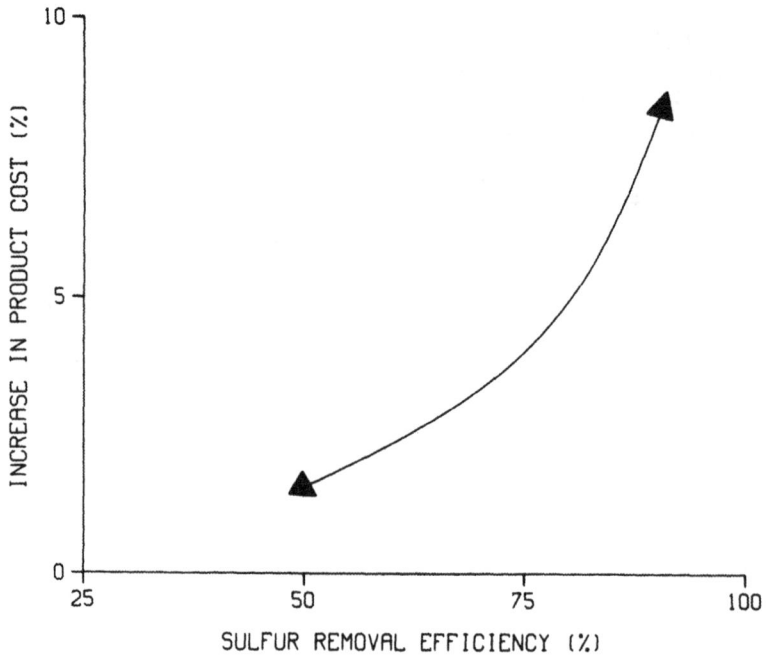

Figure 1.   Electric power SO$_2$ control costs ⌈based on $.03/kWh cost from Hesketh (10)⌉.

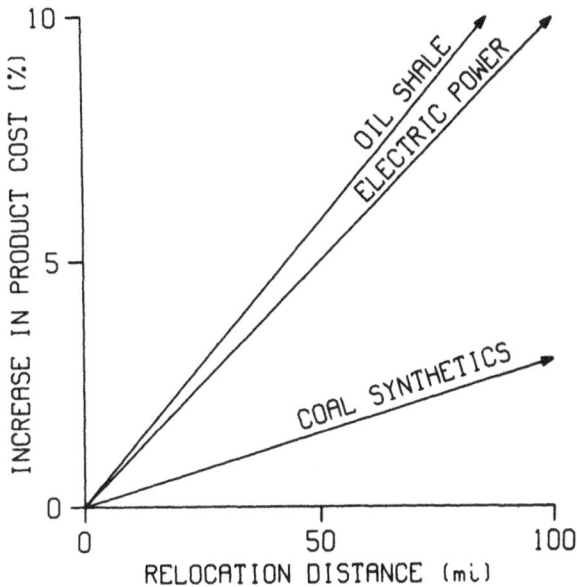

Basis

1. Coal/Shale shipment at $.02/ton-mi

2. Coal at 20 million Btu/ton

3. Electric power

   Transmission at .02 mills/kWh/mi
   Conversion at 9500 Btu/kWh
   Product at $.03/kWh

4. Coal synthetics

   Conversion efficiency at 60%
   Product at $5.25/million Btu

5. Oil Shale

   Feed at 25 gal. oil/ton
   Product at $30/bbl

Figure 2. Potential increase in costs from relocation.

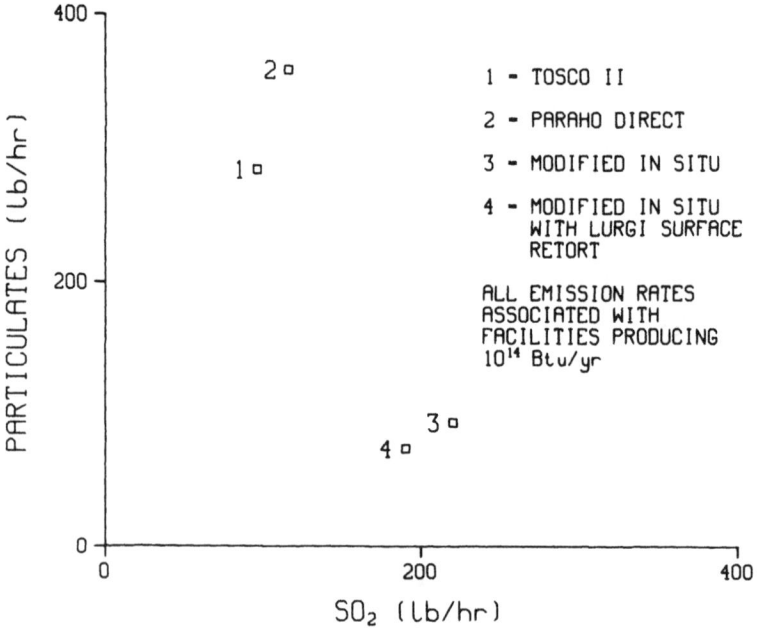

Figure 3.   Oil shale, SO$_2$ and particulate emissions.

equipment to production processes that had their genesis when pollution was not a regulated problem. Rather, one would expect that, over a period of time, inherently clean processes will be developed, to lower the cost of emitting pollutants at low levels. The effects of process changes are seen clearly in the synfuels industry. A number of processes produce oil shale. Because of fundamental differences among the processes, each yields different emissions at different costs.

Unfortunately, data in this area is skimpy; however, one analysis (13) attempted to quantify the costs of pollution control of four different oil-shale processes. Data from this source on emissions of particulates and SO_2 are shown in Figure 3. Dramatic differences in emissions of these pollutants result from the four different processes. The TOSCO process produces less of both pollutants than does the Paraho process and thus might be preferred. (The Paraho process, however, produces less NO_x than does TOSCO). The Modified In Situ with Surface Retort and TOSCO are the preferred processes. When particulate matter is especially damaging, the former process is desired; and when SO_2 is most damaging, TOSCO is preferred. When air pollution policies throw up formidable obstacles to one production process, the efficient industry response may be to move away from that process towards a less polluting one.

As with the case of synfuels, there are process change options open to electric utilities. The simplest of these is switching to higher quality fuel (such as low ash/low sulfur coal) to reduce emissions. Because most coal in the West is already low sulfur, this option may have little use in the West. The potential for fuel switching in the East and Midwest, however, is significant. Other process changes include moving to other generation processes such as combined cycle/low Btu coal gasification or fluidized bed generation. These produce very low pollutant emissions at moderate increases in costs over current coal-fired steam generation.

Demand Shifts. An actual shift in consumer demand from products of polluting industries is unlikely to have an appreciable effect on energy production in the West. The small electric power price increases created by environmental control may dampen demand for electricity, but the effect is likely to be small. For synfuels, product prices currently are set by foreign oil. Unless synfuels production is only marginally economic and environmental controls tip the balance against synfuels, there will be essentially no price effects from additional control.

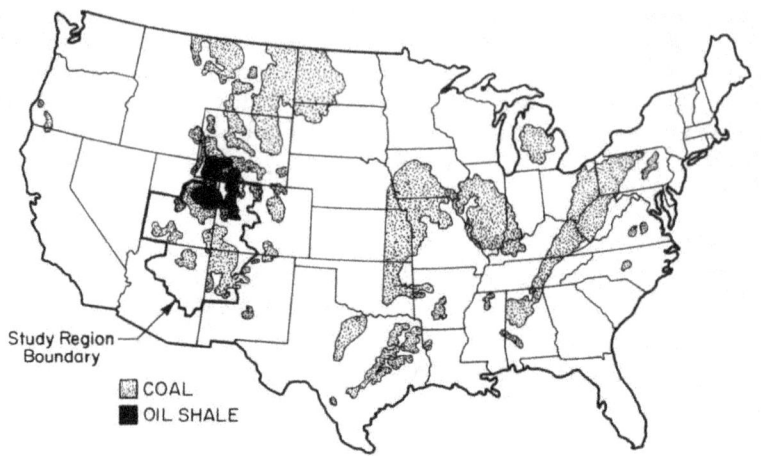

(a)   Coal and principal oil-shale regions.

(b)   Mandatory Class I areas.

Figure 4.   The Four Corners study region.

Regional Costs of Pollution Control

In the previous section we discussed the costs to individual facilities and the range of action open to industry and consumers to reduce pollutant emissions. In this section we are concerned with a region of the West with the potential for a variety of energy producers and pollution sources. In a region such as this, what are the overall costs of air pollution restrictions that allow industry to choose the best mix of strategies to reduce pollution at least cost?

We examine the Four Corners region of the Southwest. The region contains a large number of national parks and wilderness areas, those locales most highly protected by the Clean Air Act (Fig. 4a). The region also contains most of the nation's high-grade oil shale resources and significant amounts of coal for energy consumers in the Southwest and California (Fig. 4b). It is unlikely that there is any place in the US with more of a potential conflict between the goals of air quality protection and energy development.

Our approach to analyzing the costs of air pollution control in this region has been to develop a model of industrial response to air pollution regulation (14, 15). This Four Corners Model allows us to determine the total costs to industry of meeting a particular air pollution regulation, given assumptions about energy demand. For our purposes here, we will postulate three levels of energy demand for the year 1995--low, medium, and high--and then look at the average cost of supplying that energy. The total cost of meeting a particular regulation will be the difference in total cost under regulation vs without air quality regulation. Table 1 illustrates the level of demand assumed under the three scenarios. We assume that the hypothesized regulations apply to the 1987 to 1995 period.

Now let us consider the extra cost associated with controlling emissions of SO_2, one of the most significant pollutants in the study region. As we have seen, aggregate SO_2 emissions are closely correlated with regional visibility impairment. Without additional pollution controls (16), an average over the three scenarios of nearly 1.7 million tons of SO_2 is projected to be emitted annually in 1995 in the study region. What is the least cost way of reducing overall SO_2 emissions in the study region? The right-hand axis in Fig. 5 shows the total additional cost associated with reducing overall emissions. Costs presented in the figure represent an average over the three scenarios. For perspective, note that in Table 1 the average

Table 1. Four Corners region potential energy supply

	Low		Medium		High	
	1995	1987–1995	1995	1987–1995	1995	1987–1995
Electricity (10^9 kWh/yr)	82	15	140	58	177	62
Coal Synthetics (10^{12} Btu/yr)			350	350	1600	1300
Oil shale/Tar sands (10^{12} Btu/yr)			930	860	2100	1800
Total Value[a] ($\$10^9$/yr)	2.5	0.44	11	8	25	18

[a]Based on 3¢/kWh electricity, $\$5.25/10^6$ Btu synfuels

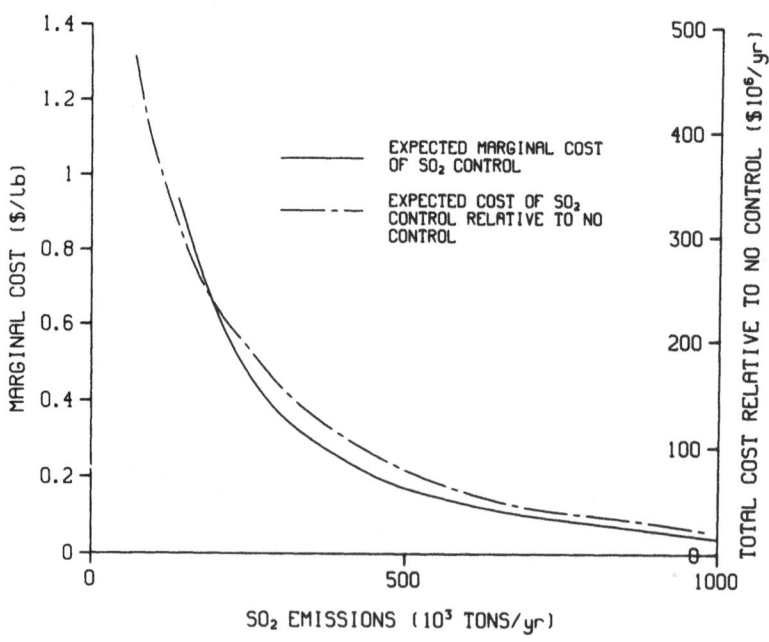

Figure 5. Costs of SO_2 control in study region.

value of the energy produced from facilities coming on line in the 1987 to 1995 period is about \$13 billion (17). A very significant reduction in SO_2 emissions could be achieved before control costs amount to even a few percent of the total value of the energy produced. A caveat must be made that Fig. 5 presents the least-cost way of achieving the given aggregate emission levels. Improperly designed regulations may not result in the moderate costs presented in the figure.

A discussion of the costs of pollution control would not be complete without a discussion of the costs of current Clean Air Act regulations. In our discussion above, we were concerned with the cost of reducing the overall level of SO_2 emissions. Current regulations are concerned with a much more complex set of air quality goals. Nevertheless, we can use the Four Corners Model to simulate the costs of current PSD regulations. Our analyses show that, averaged over the three scenarios, the extra cost of current regulations is about \$410 million annually. This translates to a cost of just under 4% of the value of the energy produced from the facilities coming on line from 1987 to 1995. Certainly the cost is significant (nearly half a billion dollars a year). How acceptable the cost is to society must be answered by society as a whole.

The SO_2 emissions from current regulations should be about 270 kilotons of SO_2 per year in 1995. As can be seen from Fig. 5, such an aggregate level of emission could be achieved at about half the cost. This however, is an unfair comparison. Current regulations protect local as well as regional air quality. A cap on regional emissions of approximately 120 kilotons of SO_2 in 1995 would provide the same level of local air quality (and significantly better regional air quality) as current regulations at slightly higher cost. Whereas it is probably possible to change current regulations to provide the same level of air quality protection at lower cost, current regulations do not appear to be dramatically inefficient, given the act's definition of air quality.

Does the act actually prevent any sources from locating in the study region? Our analysis, including considerations of visibility effects, suggests that this does not occur although siting is prevented in the vicinity of national parks and wilderness areas.

Are there other, non-energy, pollution sources, which may be cheaper to control than energy sources? Copper smelters contribute significantly to overall SO_2 loadings on the region and thus are major contributors to regional, as opposed to local, pollution. Smelters have relatively large SO_2 emissions that could be abated at costs somewhat lower than those of power plants. Existing smelters have SO_2 control efficiencies ranging from near zero to 90% or more. Some of the facilities with less effective controls could be modified at relatively low cost per pound of sulfur emitted. For example, one smelter has virtually no control on it at present, and 70% control would eliminate about 280 000 tons/yr of SO_2 emissions at an average cost of $11.6/ton. At 97% control, an additional 130 000 tons/yr of SO_2 emissions would be eliminated at an increment of $84/ton. Smelters with some control could be upgraded at a somewhat higher cost. One facility, for example, could upgrade controls from 65% to 92% and eliminate 130 000 tons/yr of emissions at a cost of $132 per ton of SO_2. Compare these costs to the marginal cost of SO_2 control from Fig. 5. The figure indicates that projected 1985 emissions under current regulations (270 kilotons per year) imply a marginal SO_2 control cost of approximately $800/ton. Thus, smelter control is a more cost effective way of reducing SO_2 loadings than strict control of energy facilities.

The Benefits of
Air Pollution Control

The benefits of air pollution control for energy related industrial facilities in the Four Corners states and the Rocky Mountains are very different from those in the eastern US. The principal economic justification for air pollution control efforts to date has been health effects. For example, Lave and Seskin (18) have calculated that the urban health benefits of SO_2 control exceed the costs of control. However, in the West, energy related industrial facilities are typically located in sparsely populated rural areas where pollutant emissions into the air are unlikely to affect significant populations.

Health benefits are typically calculated using the following formula: (Value of Safety in Dollars) x (Reduced Risk from Pollution Control) x (Population at Risk).

Obviously, if the dollar value of safety and the magnitude of reduced health risks are controversial or questioned, if the population at risk is negligible, health benefits of air pollution control will be negligible. Thus, for rural industrial facilities in the West, we are forced to dismiss health benefits of air pollution control as insignificant.

This, however, is not to say that other sources of benefits may not be extremely large. In fact, evidence is accumulating that the aesthetic effects of air pollution, in particular visibility, may be equally as important as health effects on a national scale. We briefly summarize four available studies of the benefits of preventing visibility degradation. Note that, in sum, the studies suggest that pristine visibility in the national parklands of the West may be worth more to the nation as a whole than any other single source of air pollution control benefits yet identified.

Study 1. Visibility in the Four Corners Region

The Four Corners study (19, 20) represented the first empirical attempt to value enviromental effects on energy development in the West. The roots of the effort are in Davis (21) and Bohm (22). The study investigated the impacts of Navajo coal strip mine and the Four Corners electric generating plants in the southwest region. Aesthetic benefits of abatement of environmental damage resulting from air pollution (visibility), power lines, and land disturbance from mining activities were estimated. To prevent these effects individuals were willing to pay more than $80 per year (see Table 2). No bias tests (that is, hypothetical, information, instrument, interviewer, or non-respondent sampling) were formally reported.

Study 2. Visibility and Aesthetics at Lake Powell

Lake Powell, with annual visitation now approaching two million visitor days, is an excellent example of the tradeoff between preservation and development. The lake, formed by the filling of the Glen Canyon, retains the steep cliffs, rugged terrain features, and scenic vistas one associates with the Grand Canyon. Lake Powell is now accessible to pleasure boaters and others. Construction of the Navajo coal-fired generating station located at the southern end of Lake Powell was completed in 1976. Another power plant, the Kaiparowitz Project, was also proposed for construction near Lake Powell and became an issue of substantial public concern.

As part of the Lake Powell experiment, during the summer of 1974, interviewers attempted to determine the aggregate willingness of Lake Powell users to pay to prevent construction of the proposed Kaiparowitz plant (5). Interviewers showed photographs of the existing Navajo power plant with visible pollution emanating from the stacks and with the stacks alone. All of the interviewees had seen

Table 2. Comparison of results for southwest visibility studies[a]

Non-Market Valuation Studies	Public Good	Instrument	Comparisons			
			1[b]	2[b]	3[c]	4[c]
Four Corners experiment	Visibility, soil banks, transmission lines (aesthetics of the above)	Sales tax	$85[d] (4.31)[g]	$50 (3.02)	NA[e]	$1.79[f] (0.19)
Lake Powell experiment	Visibility (aesthetics only)	Access fee	NA	NA	$2.95[h] (2.0)	$1.52 (0.29)
Farmington experiment	Visibility (aesthetics only)	Utility bills or wage tax	$82 (9.10)	$57 (4.63)	$2.44[i] (0.23)	NA

[a]The Four Corners experiment (19, 20) and the Lake Powell experiment (5) only obtained equivalent surplus bids, thus comparisons between studies are limited to subsamples of the data sets from each study.

[b]Yearly mean bids.

[c]Bid per day.

[d]The comparison between the Four Corners experiment and the Farmington experiment (23, 24) is for two alternative levels of environmental quality changes.

[e]NA--No comparison can be constructed.

[f]The comparison between the Four Corners experiment and the Lake Powell experiment required different comparisons than did the Four Corners experiment with the Farmington experiment.

[g]Standard errors in ( ).

[h]Adjusted for 6.6% inflation.

[i]Mean bid for $1.00 starting points in the Farmington experiment, which is the starting point used in the Lake Powell experiment.

the actual stacks, which remain visible more than 20 miles up the lake. Interviewers then asked what entrance fee persons would willingly pay to prevent construction of another similar plant, first, where only pollution would be visible from the lake itself, and second, where both stacks and pollution would be visible.

The analysis of the data focused on strategic bias. If users believed a uniform entrance fee might be based on the average bid of the sample to prevent construction or if users believed construction plans might be affected by the research results, then "environmentalists" might well bid very high, and "developers" might well bid zero dollars in an attempt to bias the results. A theoretical model of strategic bias was constructed to explain the distribution of observed bids likely to be bimodal rather than normally distributed if strategic bias were present. The fact that the actual distribution of bids was normally distributed was taken as evidence that strategic bias was not present. It was suggested by Brookshire et al. (5) that the absence of strategic bias might be due to the hypothetical nature of the experiment--few respondents felt that their answers would affect real world outcomes. In sampling that was randomly conducted for the four principal users of Lake Powell, on the lake, in campgrounds, at motels, and in the town of Page, the highest refusal rate was less than 1 per cent.

The average bid per family or group was $2.77 in additional entrance fees in 1974 dollars, and the total annual bid--which can be interpreted as an aggregate marginal willingness to pay to prevent one additional power plant near Lake Powell--was over $700 000. The results show impressive consistencies with the one previous study (20) in the region as well as with the succeeding Farmington experiment.

Study 3. Visibility and Aesthetics at Farmington

This study reported in Blank et al. (23) and Rowe et al. (24) attempted to establish the economic value of visibility over long distances for Farmington residents and users of Navajo Reservoir. Clearly, the ability to observe long distances is almost a pure public good. Examined in this study was the extent of certain biases that the Brookshire et al. (5) study identified. These were information, strategic, starting point, and instrument biases on compensating and equivalent surplus measures of consumer surplus.

Visitors and residents in the Four Corners region of New Mexico and Arizona were interviewed. The interviewee was shown a set of pictures depicting visible ranges. Picture set C had a visible range of 25 miles, and picture sets B and A were 50 and 75 miles, respectively. The pictures represented views in different directions from the same location--the San Juan Mountains and Shiprock.

The first part of the experiment was a bidding game, structurally similar to that of Randall et al. (19, 20) and Brookshire et al. (5). A sequence of questions on maximum willingness to pay and minimum compensation was asked through a survey instrument. The second method followed that of Rosen (25), Muellbauer (26), and Hori (27) in attempting to use the household production function, in a methodological cross check by collecting market type information through a survey instrument. The contingent behavior component of the questionnaire attempted, through contingent changes in time allocation, to infer an expenditure function and compensated demand curve, primarily by postulating an exact form of a utility function and estimating a time-related household technology (23).

It is interesting to compare results of the Farmington study with previous studies. Randall et al. (19) only reported, and Brookshire et al. (5) only obtained equivalent surplus bids. The following comparisons presented in Table 2 are, therefore, limited to the equivalent surplus bids. Using the sales tax as the instrument, Randall et al. (19) reported yearly mean bids of $85.00 ($4.31) for moves from the highest level of environmental damage, situation (A), to situation (C) representing lowest levels of environmental damage; situation (B) represented an intermediate level of damage (28). A yearly mean bid of $50.00 ($3.02) per household was reported for moves from situation (B) to situation (C). The Farmington experiment yearly mean bids for the comparable situations were $82.20 ($9.10) and $57.00 ($4.63). If one considers that the Randall et al. (20) figures should be higher as respondents are also bidding on soil banks and transmission lines, these figures are comparable.

The overall mean for situation (A), good visibility, to (C), poor visibility, in the Lake Powell experiment (5), was $2.77 ($.19) per day. Adjusted for the 6.6 per cent inflation between the time periods of the studies, these values become $2.95 ($.20). The overall mean for users of recreational sites for the comparable situation in the Farmington experiment was $4.06 ($1.11), which is considerably different. However, the mean bid was $2.44 ($1.23)

when $1.00 starting bids were used in the Farmington exper-
iment, which corresponds to the Lake Powell starting bid.
Thus, while still statistically different for the same
starting bids, the results are much closer. The Farmington
experiment, while not designed as a replication, demonstrated reasonable consistency with other studies.
Finally, a comparison of values for similar sub-samples
between the Four Corners and the Lake Powell experiments--
respectively, of $1.79 ($.19) and $1.52 ($.29)--also
suggest consistency.

Study 4. Visibility in the National Parklands

This study was designed to measure the economic value
of preserving visibility in the national parklands of the
Southwest. During the summer of 1980, over 600 people in
Denver, Los Angeles, Albuquerque, and Chicago were shown
sets of photographs depicting five levels of regional visibility (haze) in Mesa Verde, Zion, and Grand Canyon
National Parks. Although calculations in the study suggest
that projected emissions with existing and currently planned SO_2 controls would not produce a perceived decline in
visibility, complete decontrol of SO_2 emissions by projected power plants in the region in 1990 would result in a
decrease in typical summer visibility from what was represented in the photographs as "average" visibility to what
was represented as "below average" visiblity.

On the basis of this, the survey participants were
asked how much they would be willing to pay in higher electric utility bills to preserve the current average
condition--middle picture--rather than allow visibility to
deteriorate, on the average, to the next worse condition as
represented in the photographs (an estimate of total preservation value). They were also asked about their willingness to pay in the form of higher monthly electric power
bills to prevent visual plume in a pristine area. To
represent plume blight, two photographs were taken from
Grand Canyon National Park, one with a visible plume. The
surveying has a very high response rate (few refusals).

Individual household bids ranged from an average of
$3.72 per month in Denver to $9.00 per month in Chicago for
preserving visibility at the Grand Canyon. This visibility
degradation corresponds to an increase in fine particulate
matter concentrations of .4 g/m^3--a decrease in visual
range from 240 km to 210 km. These average bids were increased by $2.89 to $7.10 per month per household in the
four cities if visibility preservation were to be extended
to the Grand Canyon region as a whole as represented by the

photographs taken from Mesa Verde and Zion. Prevention of a visible plume at the Grand Canyon was worth on the average between $2.84 and $4.32 per month for interviewees in the four cities surveyed.

Extrapolating these bids to the nation implies that preserving visibility in the Grand Canyon region is worth almost 6 billion dollars per year. This is the base figure from which the benefits of power plant SO_2 controls, projected to be in place in the region in 1990, are determined. Adjusting this number for 1990 population levels and using a 10 per cent discount rate over a thirty-year power plant life gives an annualized value of 7.6 billion dollars as the benefit of power plant SO_2 control in 1990. These figures imply a marginal valuation on SO_2 of nearly $13 per pound of SO_2.

Reconciling the Costs and Benefits of Air Pollution Control

The costs of air pollution control and damage from air pollution were treated separately in the last two sections. The purpose of this section is to reconcile costs of control with damage to answer specific questions. What level of pollution control represents an appropriate balancing of costs and damages? Are current regulations consistent with such a level, or do current regulations over- or under-control pollution?

The first question concerns an optimal balancing of costs and damages. Let us focus on visibility protection, which, as we saw in previous sections, is one of the most important air quality values in the West. In terms of visibility, regional haze appears to be a more serious problem than plume blight, at least in control costs. As long as sources can be precluded from locating near Class I areas, plume blight should be a manageable problem. Regional haze control is potentially much more formidable because changing source location does not appear to be an effective control. Regional haze appears to be tied closely to aggregate emissions of SO_2 over a wide region (4, 29).

To control regional haze, suppose we place a cap on emissions of SO_2 in the Four Corners study region. We allow different sources to negotiate or trade rights among themselves so that the cap on emissions is achieved in a least cost manner. The economic model of air pollution regulation mentioned above can then be used to simulate industrial response and costs of such a lid on SO_2 emissions. In fact, the results of utilizing such a cap were presented in Fig. 5. There we showed the average (over the three demand scenarios) additional and marginal cost of SO_2 control as the lid on SO_2 emissions becomes tighter.

The optimal level of SO_2 emissions is that level where the marginal cost of emission control equals the marginal damage from a unit of emission. In the last section we saw that marginal visibility damage in the Grand Canyon region is approximately $13 per pound of SO_2. As can be inferred from Fig. 5, at an optimum this represents a very low level of SO_2 emissions, far lower than that implied by current regulations. Unquestionably there is a great deal of uncertainty in these damage estimates. However, even if they are an order of magnitude too high, they suggest that current regulations are not nearly strict enough to control regional haze in the study region.

Conclusions

The current system for regulating air pollution has a number of important features that relate to its efficiency in developing energy while preserving air quality. First, the Clean Air Act permits large development in the mountain West. Second, by requiring best available control technology, the act appears to be relatively effective at containing growth of SO_2, currently the pollutant with emissions most responsible for the visibility impacts in the region. Third, the act encourages siting away from Class I areas and thus diminishes the likelihood of visible plume impacts in national parks. Finally, the act appears to achieve these goals at costs commensurate with other regulatory alternatives as long as only energy sources are considered.

Our review of the willingness to pay to avoid visibility degradation in the Grand Canyon suggests that visibility is a highly valued resource in the west. Further, given the visibility protection provided by current regulations, SO_2 regulation is not overly strict. In fact, current regulations may undercontrol SO_2 in relation to visibility degradation.

However, the act does have some shortcomings that may prove increasingly important in the future. First, control of oxides of nitrogen is relatively indirect. Nitrates formed from oxides of nitrogen are apt to exhibit a threshold phenomenon with respect to visibility. At low concentrations and high temperature, nitrates are gases and consequently have negligible effects on visibility. At higher concentration with lower temperatures, small nitrate aerosols will form and influence visibility.

Similarly, in the instances where increments are met by efficient control of sulfur oxides, the visibility protection against plume blight may be inadequate. In these instances, the Federal land managers must demonstrate adverse effects, and budget pressures may prevent employment of sufficient resources to make such demonstrations.

While the act seems to be reasonably efficient at achieving air quality goals if sources with similar costs are considered, it may be less so if other sources are considered. Retrofit of SO_2 controls for smelters offers two options. First, it would be possible to achieve the same overall air quality at lower cost if smelters were controlled and lower controls were required of power plants. Second, significantly improved air quality in the parks could be achieved if smelters were controlled in addition to effective controls on power plants. The very large damage estimates associated with visibility impairment appear to imply that further control of SO_2 is warranted.

The act also has a major shortcoming in the regulation of fugitive dust. The construction of a new plant presently may result in non-attainment of ambient standards with negligible emissions. To date, fugitive dust has not played a major role in plant siting or control, but in the future such regulation could be a significant factor in plant siting decisions.

References and Notes

1. J. Trijonis, "Visibility in the Southwest--The Exploration of the Historical Data Base," Atmos. Environ. <u>13</u>, 833-843 (1979).

2. D. Nochumson, "An Evaluation of Regional Haze Visibility Impacts," 8th Energy Techn. Conf. Exposition Proc., Washington, DC, March 9-11, 1981.

3. "Matrix Methods to Analyze Long-Range Transport of Air Pollutants," US Department of Energy, Assistant Secretary for Environment, Office of Environmental Assessments Regional Impacts Division report DOE/EV-0127 (January 1981).

4. D. Nochumson, "NCAQ Report 2: Air Quality in the Four Corners Study Region. Vol. II: Regional Analysis," Los Alamos National Laboratory LA-UR-81-1145, March 1981.

5. D. Brookshire, B. Ives, and W. Schulze, "The Valuation of Aesthetic Preferences," J. Environ. Econ. Manage. 3, 4, 325-346 (December 1976).

6. G. R. Neuroth, "Ambient Visibility and Particulates in Arizona During the 1980 Copper Smelter Strike," Arizona Department of Health Services, Bureau of Air Quality Control (December 31, 1981).

7. C. A. Mangeng and R. Mead, "Sulfur Dioxide Emissions from Primary Non-Ferrous Smelters in the Western United States," Los Alamos Scientific Laboratory report LA-8268-MS (August 1980).

8. P. A. Hamilton et al., The Reserve Base of U.S. Coals by Sulfur Content, Volume II: The Western States, Information Circular 8693 (Department of the Interior, Bureau of Mines, Washington, DC, 1975), p. 7.

9. Ambient standards are levels that define the allowable concentrations of pollutants at ground level.

10. H. E. Hesketh, "Economic Process Technology and Cost Curve Development Data and Procedures for Coal-Fired Electrical Generation Facilities," in "Air Pollution Control Costs and Regional Impacts in the Four Corners Study Region," F. Roach and J. Jaksch, Los Alamos National Laboratory LA-UR-81-1379, Los Alamos, New Mexico (March 1981).

11. Examples of such change are numerous and include the Huntington Canyon Power Plant and Intermountain Power Project, both in Utah.

12. Actual costs may be less because a move of 100 miles may not increase transport distances by 100 miles.

13. "Predicted Costs of Environmental Controls for a Commercial Oil Shale Industry," prepared by the Denver Research Institute for the US Department of Energy COO-5107-1 (July 1979).

14. The model is a static partial equilibrium model with the Four Corners region divided into 22 subregions. For more information, refer to Roach et al. (15).

15. F. Roach, J. Bartlit, A. Kneese, C. Kolstad, D. Nochumson, R. Tobin, G. Weatherford, and M. Williams, "NCAQ Report 4: Alternative Air Quality Policy Analysis for the Four Corners Study Region," Los Alamos National Laboratory, LA-UR-81-1380, March 1981.

16. Above and beyond those installed before 1987.

17. Unless otherwise indicated, costs are in 1980$ throughout this chapter.

18. L. B. Lave and E. P. Seskin, <u>Air Pollution and Human Health</u> (Johns Hopkins Press, Baltimore, 1977).

19. A. Randall, B. Ives, and C. Eastman, "Benefits of Abating Aesthetic Environmental Damage from the Four Corners Power Plant, Fruitland, New Mexico," New Mexico State University Agricultural Experiment Station, Bulletin 618 (1974).

20. A. Randall, B. Ives, and C. Eastman, "Bidding Games for Valuation of Aesthetic Environmental Improvements," J. Environ. Econ. Manage. $\underline{1}$, 2, 132-149 (August 1974).

21. R. Davis, "Recreation Planning as an Economic Problem," Nat. Resour. J. $\underline{3}$, 239-249 (May 1963).

22. P. Bohm, "Estimating Demand for Public Goods: An Experiment," Eur. Econ. Rev. $\underline{3}$, 2, 11-130 (1972).

23. F. Blank, D. Brookshire, T. Crocker, R. D'Arge, R. Horst, Jr., and R. Rowe, "Valuation of Aesthetic Preferences: A Case Study of the Economic Value of Visibility," Resource and Environmental Economics Laboratory, University of Wyoming report to the Electric Power Research Institute, 1977.

24. R. Rowe, R. D'Arge, and D. Brookshire, "An Experiment on the Economic Value of Visibility," J. Environ. Econ. Manage. $\underline{7}$, 1, 1-19 (March 1980).

25. S. Rosen, "Hedonic Prices and Implicit Markets: Product Differentiation in Pure Competition," J. Pol. Econ. $\underline{82}$, 1, 34-55 (January/February 1974).

26. J. Muellbauer, "Household Production Theory, Quality, and the 'Hedonic Technique'," Amer. Econ. Rev. $\underline{64}$, 6, 977-994 (December 1974).

27. H. Hori, "Revealed Preferences for Public Goods," Amer. Econ. Rev. 65, 5, 978-991 (December 1975).

28. Standard errors are in parentheses.

29. M. D. Williams, C. A. Mangeng, S. Barr, and R. Lewis, "NCAQ Report 2: Air Quality in the Four Corners Study Region. Vol. I: Local Analysis," Los Alamos National Laboratory, LA-UR-81-1145, March 1981.

11. Energy Development:
A Challenge for
Environmental Planning

During the later years of the '70's, it became popular to refer to a "War Between the States," specifically the movement of people from the so-called Frostbelt to the Sunbelt. The shift in population meant several things to the states involved; reallocation of voting power, grants-in-aid, etc. The impacts and political ramifications are controversial when the Federal/state relationships change because the formulae established by law or regulation respond to stimuli (population movement) not directly caused by either party. How much more friction is possible when the involved parties also become one of the factors causing the change?

Add this potential irritant to the likelihood that any impact is most apt to become evident sometime in the future, and therefore has to be forecast. We then have a situation of one party blaming the other for possible impacts in the future, with both the blame and the impact subject to considerable debate.

One example of such a situation was the anticipated growth in the Mountain States in the decade ahead promoted by the deployment of the MX missile system and the development of synfuels industry. By far, the biggest impetus was the synthetic fuels development anticipated by the Synthetic Fuels Corporation.

Former President Carter, in his July 15, 1979, energy address, set a limit to the amount of oil this country would import at the level of 1977. He also announced the goal of cutting our dependence on foreign oil by 50 percent by 1990--a reduction of over 4.5 million barrels per day of imported oil. To help his 1990 import reduction goal, the President proposed the development of 2.5 million barrels per day of oil substitutes from coal liquids and gases, oil shale, biomass, and unconventional gas. Of this 2.5 million level, 400,000 barrels per day (BPD) was to be the shale oil contribution. This remains a national goal.

After the President's energy address, Congress acted to create the Synthetic Fuels Corporation to provide financial assistance to synthetic fuel developers. The goal of the Corporation, as outlined by the Congress, was to foster the creation of a diversified synthetic fuel industry of at least 500,000 barrels per day by 1987, and 2,000,000 barrels per day by 1992. (These goals do not include fuels from biomass and unconventional gas, which were included in the President's earlier program.)

The growth of energy in the Mountain States has caused some concern because of its topography, extensive Federal land ownership, and possible environmental, social, and water constraints. The population of the area is expected to double by the year 2030, with total employment growing by two and one-half times during the same period. This population growth, fostered and followed by economic development of all kinds, will have all kinds of repercussions. Here is a sampling (1):

Socioeconomic. If the growth in some of the rural counties of these states is as great as, say, 10 percent or more in two or more consecutive years, the socioeconomic impacts are likely to be severe. The major difficulties are financial in nature as the communities rush to put in place the infrastructures required to support their new citizens. But management and administrative skills may also be wanting.

Among the more readily noticeable impacts will be change in the rural-western lifestyle. Old traditional values are not usually amenable to urban environments. Infusion of people from elsewhere in the nation with other ideals and beliefs will hasten this change. For better or worse, the "Old West" will fade from the scene.

Transportation. The transportation system in the area is adequate for the present. It has been built largely along paths of least resistance in a rugged terrain. Large influxes of people and industry will necessitate significant expansion of roads, railroad beds, and pipelines. In code terms alone, the impacts are likely to be significant.

Land Use. The most prevalent economic activity in these states today is agriculture, which is closely tied to water resources. Although there may be technological fixes which can bring water to new acreage to replace that which goes out because of urban or industrial development, there is a question whether new land uses are compatible with the existing styles in agriculture; for example, rangeland cattle raising. In addition to changes in the number of acres devoted to agriculture over time, there may be changes in basic agriculture practices.

Urban land use in the West has traditionally tended to sprawl. To some extent, as development increases, single-family houses will be less spaciously placed but even so there is little possibility that these new Western suburban areas will take on the density patterns of Eastern cities any time in the near future. This tendency to disperse will have ramifications for communications, transportation, environment, and other areas.

Finally, many areas in the West have long been known for access to splendid outdoor recreation. Camping, hiking, fishing, skiing, hunting, and the like are not necessarily compatible with widespread development. Certainly, there will be large changes in the way people there live, work, and play.

Because this paper was planned to focus heavily on the environmental impacts of synthetic fuels development in the West, I turn to an analysis done in August 1980 for Senate hearings in Colorado (2).

It is clear from the projections that oil shale is expected to be a major contributor to the future synthetic fuels production of this country. In the recently announced assignment of awards for the Alternative Fuels Production Program, four oil shale projects were selected totalling 12.8 percent of the $100 million feasibility study effort. These feasibility studies are designed to determine whether a specific project is technologically, financially, and environmentally viable.

The Department of Energy, and its predecessor organization, the Energy Research and Development Administration, have over the years attempted to address environmental impacts of energy development in these areas. Previous studies have been inconclusive and have lead to widely varying predictions because data were not available from actual commercial operations and the prediction models used were imprecise tools. The studies do not identify any serious impediments to reaching the goal of 400,000 BPD. Thus, the initial increment of

400,000 BPD is both a safe and necessary step to achieving whatever maximum capacity is ultimately established for the oil shale industry. Similarly, no precise determination of growth rate is possible; however, the institutional constraints (permits, land and regulatory problems) for planned projects probably will limit growth to relatively low rates in the initial period of oil shale development.

We are especially concerned about the air quality, water availability, and negative socioeconomic impacts. These will be discussed greater detail.

Oil Shale: Air

A number of studies conducted since 1976 have examined the air quality impact of oil shale. At first glance, these reports appear to be enormously contradictory: some say there can be essentially no oil shale development while others predict a substantial industry can be built. A closer examination of these reports, however, suggests other conclusions. Early estimates, based on whatever information was available, tend to be conservative. As the body of information on oil shale has grown over time, these conservative assumptions have moderated, yielding generally higher projections for the size of an industry which can be sited. Another important factor is that, generally, the focus of these reports have been to identify the impact of a single plant or an assumed level of development (e.g., 200,000 barrels per day), rather than to determine the maximum level of development possible due to air quality regulatory constraints. This means there is likely to be range of estimates, none of which, however, is aimed at the maximum carrying capacity of a region. A brief review of these studies is informative.

A 1976 report for the Energy Research and Development Administration gave perhaps the lowest estimate for oil shale development when it stated: "...control levels beyond the best available technology will be needed for particulate and SO_2 emissions from synthetic liquid fuel plants (3). Using the TOSCO II process as a prototype, the study noted a single 100,000 barrel per day plant could not be sited without a significant reduction in particulate emissions, and a marginal reduction in SO_2 emissions.

DOE published studies in July 1979 and June 1980 on synthetic fuels. The 1979 study concluded that "air quality considerations would not preclude construction of judiciously sited and efficiently controlled" oil shale technologies, up to the 400,000 barrel per day Carter

Administration proposal (4). The 1980 study also concluded that the 400,000 barrel per day level was practical from an air quality perspective (5). No upper limit was determined.

Several EPA working papers have also examined the air quality impacts of oil shale development. A 1976 memorandum from David Joseph to Terry Thoem (both in EPA's Denver Office) stated a 200,000 barrel per day capacity was feasible. A November 1979 analysis based on the Colony and Union oil permit applications also identified 200,000 barrels per day as acceptable. In a paper presented in the spring of 1980 in Golden, Colorado, Mr. Thoem raised the estimate to 400,000 barrels per day, based on a "box" model (most of the other studies cited here are based on diffusion models). These limits were projected as due to PSD I areas and SO_2. A draft study prepared by EPA (IERL) in the spring of 1980 concluded that at least an 850,000 barrel per day capacity can be sited. Finally, in his July 17, 1980, testimony before Senator Hart, Mr. Thoem projected air quality constraints starting at the 1.1 million barrel per day level (due to ozone).

Based on the earlier EPA analysis (spring 1980), the Office of Technology Assessment stated that the Clean Air Act "could limit production in Colorado to 400,000 barrels per day" (6).

A summary of the outcome trend of these reports is presented in Figure 1.

One should be careful not to read too much into the apparent trend these projections present. They indeed can be seen to represent a shifting opinion as to how much capacity can be safely sited, but curve extrapolation is not justified, as explained below.

Several factors probably contribute to the increase in siting capacity projections. First, and probably most important, is the fact that in 1976, when these projections started, oil prices were too low to allow synthetic fuels to compete successfully in the market place. Therefore, interest centered on the feasibility of one or two demonstration plants, not an entire industry. Projections at that time, and to a lesser extent at later times, reflected their author's opinion that at least (not at most) a certain capacity could be sited. Secondly, technology has evolved in two directions. Controls for specific shale technologies have improved, and also inherently cleaner, second generation shale oil technologies have entered the projections. These improving

---

1976 (ERDA)

- Based on TOSCO II
- PSD Class II limits for PM
- Controls beyond RACT needed to allow a single 100,000
- BPD facility

September 1979 (EPA)

- Based on extrapolation of first permits
- Flattops PSD I SO_2 is limiting factor
- 200,000 BPD

Spring 1980 (EPA)

- Based on EPA short-range models and announced plans
- Flattops PSD II SO_2 is limiting factor
- At least 400,000 BPD

June 1980 Preliminary Study Results (EPA)*

- Based on reactive plume models and extended plans
- At least 890,000 BPD

July 17, 1980, EPS Testimony

- Extrapolated from June data and announced plans
- Ozone is first size limitor for health NAAQS
- Colorado development limits due to NAAQS are:

    - Ozone/0.9 million BPD
    - NO_x/3.1 million BPD

  due to PSD I are:

    - SO_2/1.7 million BPD
    - TSP/5.4 million BPD

---

*As reported in 27 June 1980 "Synfuels" newsletter; confirmed by 14 July 1980 telephone call from D. Carter (DOE) to T. Thoem (EPA).

Figure 1.   Summary of Air Quality Estimates

technologies, however, may not continue to evolve at the same rate. A third area of change, which is likely to continue changing with time, is modeling of air quality impact itself. The oil shale areas are typically located in irregular terrain, the most difficult modeling situation. They have been modeled differently in several studies. Development of better models will likely occur over the next few years, and this will affect the capacity deemed acceptable.

A final category of factors which have varied over time could perhaps best be termed "details." After the major decisions are made concerning what process will be modeled, what control technology will be assumed, and what model will be employed, a number of less conspicuous assumptions must be made. These included the source of power for the facility, the "cleanliness" of fuels burned on-site, the height of emission stacks, the ducting arrangements (minor air emissions streams can be combined and emitted through a single stack), the plant configuration, and terrain assumptions.

Given all of these variables, it is remarkable, I believe, that the recent projections from different government organizations have been as close as they have. Although air quality modeling is as much art as science, and is inherently inaccurate, it is the best tool we have to predict the future impacts of plants and prevent degradation of our air resources.

In summary, over the past four years, various studies have predicted that suitable oil shale capacity using best available air pollution control technology could range between less than 100,000 barrels per day to over 1 million barrels per day. The trend, with respect to time, is towards a larger projected capacity. The two primary factors in these changes are more detailed technology controls, and changes in modeling techniques and assumptions.

Water Availability

The issue of water availability for oil shale development in Colorado is a complex one because it is a function of so many different parameters. To compound the issue even further, as important as the purely physical aspects of water availability are constraints imposed by the institutions governing the allocation, storage, and distribution of water resources.

In the long term, availability is a function of average annual surface flow and groundwater that can be recharged, less that amount of water which is currently being used, demands set by lower Basin rights, and all projected future energy and non-energy demands in the Upper Colorado basins. Long-term availability is not the only consideration necessary for a determination of water availability, since oil shale facilities require a continuing water supply during low flow as well as high flow conditions.

For this reason, short-term or instantaneous supply is a function of unallocated surface flow, drawdown from impoundments and reservoirs and groundwater, less existing demands (including protection of instream flows for environmental purposes), lower basin rights, and a factor to allow for constraints--based on restrictions to inter-basin transfer and on priority of uses--on redistribution of water from impoundments.

I would now like to address each of these individual features which, taken together, will ultimately determine the net availability of water for oil shale development in Colorado. In doing this, I have drawn from studies by the Colorado Department of Natural Resources, Office of Technology Assessment, the General Accounting Office, and the Department of Energy. Other studies exist, although these are among the more recent. The most current and complete studies are those of the Colorado Department of Natural Resources. Among these Colorado studies is a draft regional assessment of the impact of emerging energy technologies--including oil shale development--on the region's water resources, done in cooperation with the Water Resources Council under the authority of section 13(a) of the Federal Non-Nuclear Energy Research and Development Act of 1974.

Water Requirements for Oil Shale

The first key factor I will discuss is water requirement for various oil shale technologies. I believe that there still exists in some quarters a common misconception that there is not enough water to support energy development in the Upper Colorado Basin. One of the probable reasons for this is that some early assessments predicted that water supplies simply were not adequate to meet projected energy requirements. A recent General Accounting Office (GAO) report (7) addressed this issue and concluded that the Upper Colorado Basin should have sufficient water for energy development until at least the year 2000. In particular, it showed that many of the assumptions used in an earlier study, the Department of the Interior's 1974 report (8), have, in retrospect, turned out to be too conservative. For example, growth rate estimates of electric power capacity in the West decreased by over 30 percent from 1974 to 1977. In addition, power plants in the Upper Colorado Basin actually consume much less water than that study estimates (8,000 to 13,000 acre-feet per 1,000 MW, rather than estimated 15,000 to 20,000 acre-feet per 1,000 MW).

Over time, similar water conserving or water-efficient methods have been applied to projected synthetic fuel technologies, lowering projected water requirements from earlier levels. These changes are even larger on a unit basis as the engineering specifications of these new technologies are developed. The 1974 Interior study (8) estimated water requirements for a 100,000 barrel per day oil shale operation at 17,400 acre-feet per year, while the Colorado DNR, in 1979 (9), estimated an 11,600 acre-feet per year average per 100,000 barrels per day production—a reduction of over 33 percent per plant. In general, estimated water requirements for specific projects have ranged from 4,900 acre-feet per year to over 10,000 acre-feet per year for a 50,000 barrel per day plant, depending on the specific oil shale technology. These included projections made in a study by the Denver Research Institute sponsored by my office and completed in 1979 (10), and often cited in other comprehensive assessments such as the Office of Technology Assessment report of June 1980 (6).

The range of water requirements for four different oil shale technologies in that study are:

Paraho Direct	5,346 acre-feet/year
TOSCO II	10,694 acre-feet/year
Modified Insitu (MIS)	5,817 acre-feet/year
MIS with Lurgi	5,656 acre-feet/year

These reductions in estimated water requirements hold true for other synthetic fuel technologies, and are a result primarily of (1) projecting more extensive use of dry cooling—which uses air rather than water—and (2) increased recycling and reuse of waste and process waters.

Therefore, the water requirements corresponding to a 500,000 barrel per day oil shale industry, including associated municipal growth, could range from less than 50,000 acre-feet/year to upwards of 100,000 acre-feet per

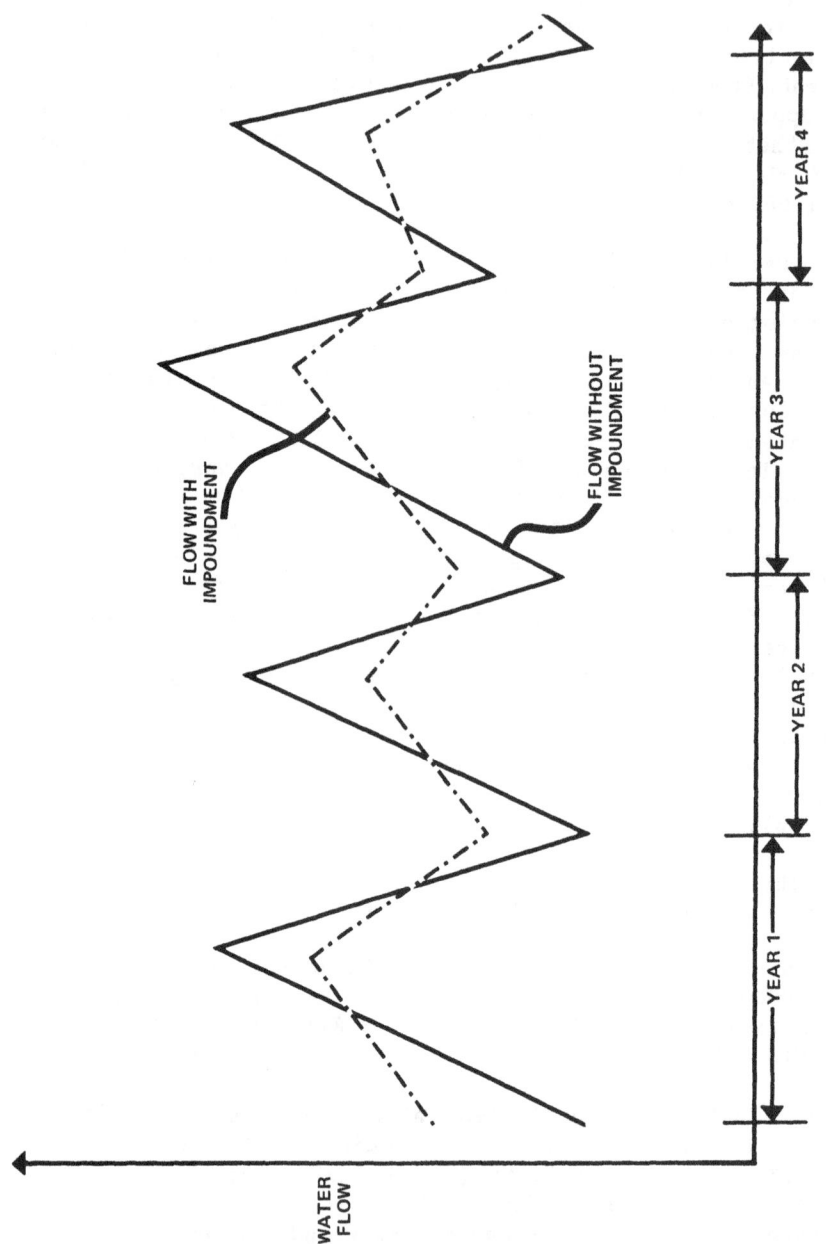

Figure 2. Impact of impoundments on river flow

year. As a comparison, current depletion by irrigated agriculture in the State of Colorado is approximately 1,200,000 acre-feet per year, out of total State of Colorado depletions of 1,800,000 acre-feet per year from the Upper Colorado River.

Water Supply

One of the reasons that water availability for oil shale development is a concern to people in this region, and to the Department of Energy as well, is that the West is a relatively arid region. Although the Colorado River Basin is large in area, its flow is relatively small, averaging 13.8 million acre-feet between 1930 and 1974, compared, for example, with the much greater flows of the Columbia and Mississippi Rivers of 180 million and 440 million acre-feet per year, respectively.

In addition, flows vary annually and seasonally, being quite high during the spring snowmelt and extremely low in dry months. These flow variations, which become most critical when the low flow cannot meet all needs, have necessitated the construction of reservoirs or impoundments, to mitigate or dampen seasonal fluctuations, and augment natural supplies during dry years. Figure 2 illustrates the effect impoundment has on seasonal flow variations, especially with respect to raising the level of low flows with captured spring runoff.

A number of reservoirs have been built in the region by the Water and Power Resouces Service (WPRS), formerly the U.S. Bureau of Reclamation. Table 1, from the GAO report (11), lists some of these reservoirs in energy resource areas which could supply future water allocations to oil shale and other energy developments with their industrial allocations. In addition, the recent report by the Colorado Department of Natural Resources (9), indicated that the Green Mountain Reservoir may have 90,000 acre-feet per year available for sale.

The Colorado Department of Natural Resources (DNR) has recently completed a draft assessment of water-for-energy for the Water Resources Council, with support by the Department of Energy pursuant to section 13(a) of the Federal Non-Nuclear Research and Development Act (12). Late in 1981, the Water Resources Council was scheduled to publish a condensation of the draft Colorado DNR report along with an assessment of the Lower Basin effects of energy development in the Upper Basin. Colorado's draft report addressed a hypothetical baseline scenario in the

TABLE 1. Selected Reclamation Reservoirs Located in Energy and Mineral Development Areas.

Reservoir	Basin	Long-Term Industrial Allocation (acre-ft/yr)
Ruedi	Upper Colorado (Frying Pan River)	47,000
Fontannele	Upper Colorado (Green River)	228,000
Navajo	Upper Colorado (San Juan River)	115,250
Powell	Upper Colorado (Colorado River)	142,000
Total		525,000

Source: (11)

TABLE 2. Water rights held by oil shale developers.

Existing Rights	Estimated Quantity (acre-feet/year)
Colony	171,000
Union Oil	85,000
SOHIO	72,000
TOSCO	39,000
MOBIL	36,000
Superior	17,000
TOTAL	420,000

Source: (13)

year 2000 of 1.3 million barrels per day of oil shale,
825,000 of which was sited in Colorado, and eight coal
gasification plants (or a total of 1.5 million barrels per
day liquids-equivalent). It concluded that the water
demands of such an industry and its associated growth could
be satisified from surface supplies without having to
impact other projected consumptive uses in the Upper Basin.

This conclusion, however, is based upon several
assumptions: that water is purchased from new or existing
reservoirs and/or that new pipeline and pumping facilities
are built to transport the water, and that no reductions in
water available for use in the Upper Basin are required to
guarantee needed flows in the Lower Colorado Basin, or to
meet Mexican treaty requirements. The study estimated the
total capitalized cost of such facilities at $1 billion for
a 1.5 million barrel per day industry, or 1 to 2 percent of
the capitalized costs of the energy facilities proper.
Most of this additional reservoir capacity would be located
in the White River Basin.

The study also addressed an accelerated level of oil
shale development of 2,400,000 barrels per day, 1,700,000
of which was in Colorado. The body of the DNR report
indicated that even this level of development might be
accommodated with greater storage development and
associated costs.

In addition, the DNR projected depletions by
non-emerging energy technology uses to increase from the
current 3.8 million acre-feet to nearly 5.4 million
acre-feet in the year 2000. To allow for protection of
present users in the Upper Basin, several assumptions were
made by the DNR, including a 500 percent increase in
consumption by steam-electric power; no transfers of water
rights from irrigated agriculture to oil shale; and all oil
shale developers' water rights are junior to all other
rights. The DNR draft report also assumed the construction
of all currently authorized irrigation projects by 2000,
although each project will, of course, have to be evaluated
on its individual merits prior to any Federal decision to
begin construction.

Water Rights

As is generally known, water rights in Western States
are based upon the prior appropriation doctrine, under
which applications for the right to use water are filed
with state engineers or water courts. In the Upper
Colorado Basin, conditional water rights, many of which

have not yet been put to use, far exceed the actual flow of the Colorado. Unallocated surface water, however, is available in some of the sub-regions, as indicated by the Colorado DNR draft assessment (12).

Many potential oil shale developers have acquired relatively large holdings of water rights, either in the form of conditional decrees or purchased irrigation rights. Table 2 has been prepared from material gathered for the 1979 Environmental Protection Agency study, "Energy from the West" (13), and shows estimated existing conditional water rights held by oil shale developers.

Other options for obtaining water are purchase of surplus water from existing reservoirs and construction of new reservoirs; interbasin transfers, which may be constrained by institutional problems and legal restrictions; and the development of groundwater supplies.

Interbasin Transfers

Interbasin transfers, or diversions of water from one major basin to another, have often been mentioned or proposed as alternatives to increase local water supply. Indeed, diversions from the Colorado and Upper Missouri River Basins to the Upper Colorado have specifically been pointed out as potential suppliers of additional water to the oil shale area.

Although the financial and physical feasibility of interbasin transfer seem viable, a number of institutional factors present some rather formidable obstacles. First, there is a Congressionally-mandated moratorium (first required by the Colorado River Basin Project Act in 1968) which extends to 1988 on any reconnaissance studies by the Department of the Interior of any plan for the importation of water into the Colorado River Basin.

Independent of the moratorium, there are two river compacts which incorporate provisions which would make it difficult to export water into the Colorado system. Both the Snake and Yellowstone River Compacts do not allow out-of-basin transfers without approval of the signatory states. No such compact provisions exist on the Upper Missouri.

Intra-basin diversions, or transferring water from one sub-basin in the Upper Colorado system to another (e.g., the Green River Basin to the White River Basin) are at least institutionally feasible. Such a diversion would be subject to provisions of the Upper Colorado compact, especially with respect to state shares of river flow, and to prevailing conditions regarding existing water rights.

Groundwater

It would be worthwhile to discuss the potential for industrial use of groundwater in more detail. Most water availability assessments, including the oil shale assessment by the Colorado DNR (12), have limited the available water supply options to surface water. This is because, in general, not as much is known about groundwater as surface flows, which can be easily measured. The Colorado DNR assessment pointed out that estimates of the amount of groundwater in storage in the Piceance Basin range from 2.5 million to 25 million acre-feet. The amount of water which is discharged to surface streams, and recharged primarily of snowmelt, is estimated at from 26,000 to 29,000 acre-feet per year.

If only an amount equal to the recharge rate in the Piceance Basin were developed, it could support as much as 250,000 barrels per day of oil shale development. In fact, mine drainage water in the central portion of the Piceance Basin where the current Federal lease tracts are located could produce from 6,400 to 18,000 acre-feet per year for those two tracts alone.

The Office of Technology Assessment report, "An Assessment of Oil Shale Technologies", (6) indicated that much of the stored water in the Piceance Basin is of such low quality (being high in dissolved solids and fluorides) that its use for non-energy purposes would not likely increase. It further estimated that if oil shale developers upgraded this groundwater, and used 15 percent of the maximum 25 million acre-feet, the aquifers alone could supply a 1 million barrel per day industry for 20 years. This rate of consumption would, however, exceed the recharge rate, and result in reduced surface flows in some streams in the Piceance Basin. The Colorado DNR assessment (12) indicates that groundwater development would produce much less adverse impact on fish and recreation than relying on surface water use alone.

Summary

In summary, it appears that sufficient options exist for developing the water necessary to support an oil shale industry in Colorado. Uncertainties related to each of these options make it difficult at this time to provide the

detailed identification of the size of an industry that can
be supported by available water resources.

Socioeconomic Impacts
of Oil Shale Production

A common theme is evident within all the discussion
presented here: that we are all vitally concerned with the
identification and management of environmental impacts
associated with oil shale production. Analytical efforts
in two of these areas, air quality and water availability,
are well along toward specific identification and
resolution of the magnitude of the problem. The third,
direct and indirect socioeconomic impact, is not as
well-defined in terms of the scope of the problem or
mitigation requirements. Air and water impacts have been
assessed using detailed deployment patterns and engineering
specifications, while the financial and management
initiatives required to provide adequate social service
support need more in-depth analysis of the impacts related
to the existing and new in-migrating population.

Types of Socioeconomic Impact

There are two types of socio-psychological in nature.
Rapid growth of rural areas, especially when due to the
introduction of high-technology industries, may alter the
way in which residents view their community and its
characteristics. Often, such growth means changes in the
way residents interact with one another, decreases the
levels of social control and quality of life, and growth
often creates conflicts between new and long time
residents.   Social problems such as divorce, crime, drug
abuse, and mental illness usually accompany such growth.
These socio-psychological impacts are not simply associated
with the construction phase of the new industry, but rather
are problems associated with the transition to and
maintenance of a growing urban area.

The second impact is a more direct result of rapid
growth. It is the inability of local and state governments
to design and pay for an infrastructure which adequately
supports the rapidly growing population. This problem is
especially acute during the initial stages before adequate
public revenues are generated by the operation of the new
facilities. Let me provide you some data which demonstrate
the magnitude of the problem. According to the oil shale
technology study prepared by the Office of Technology
Assessment (6), the population level for the Piceance Basin
associated with a 500,000 barrel per day oil shale industry

in 1990 is 280,000; this is 107,000 or 60 percent beyond
the normal growth for the area expected separate from the
accelerated energy technology development.

Both of these socioeconomic problems have common
elements. Both have a related economic demand placed upon
the local public sector. In both cases, up-front capital
facilities are required within the geographic area, an area
which currently has a minimum infrastructure support
capability because of its rural nature. The population and
economic behavior about which we are talking is not the
classic "boom-bust" situation which has so often occurred
in rural areas where a single large plant was added to a
small community. Rather, it is a situation where we can
expect rapid, but continuing, urbanization, with the
industry base oriented toward expanding and long-time use
of the area resources. These socioeconomic impact problems
are therefore cash flow problems associated with an initial
low capital base availability to the local government. The
ultimate concerns in the socioeconomic impacts area are
what quantity and quality of infrastructure support should
be provided to the growing population and on what
schedule? What is the associated cost and how can this
cost be met?

The Impact Assessment Method

The assessment method used to answer these questions
requires estimates of variables related to population
growth, employment, and area characteristics. The
principal variables for these assessments is the technology
mix and social deployment for the oil shale demonstration
and production facilities. In this way, we can define the
labor force required to construct and operates these
facilities. Based upoon the work force level, indirect
employment can be estimated. Also associated with work
force is an additional population sector which represents
the families of these laborers. Once these numbers are
estimated, usually through the use of appropriate
multipliers, the level, timing, and cost of the required
infrastructure can be estimated. These infrastructure
costs can be met through direct revenue, subsidization,
taxiation, and/or bond sales. The relationship between
these various elements is demonstrated in Figure 3.

The assessment method mentioned above is also
influenced by the character of the geographic area within
which the oil shale production industry is located. The
size and make-up of the historic labor force will influence

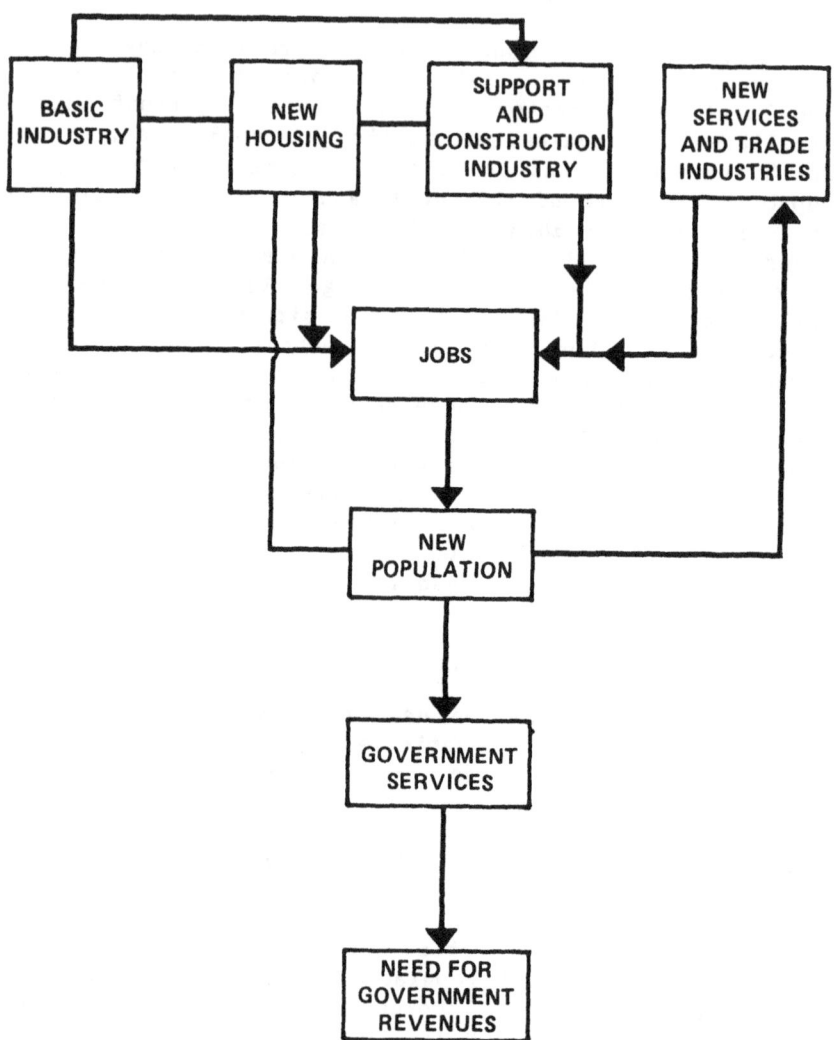

Figure 3.   Relationship of infrastructure assessment elements

the level of indirect employment which will occur and that must be met by new in-migrating workers. The physical topography of the area will affect the residential patterns of the growing population of new worker families while the economic base of the area will also affect secondary employment generation and growth rate beyond initial unused capacity. Finally, the level of the existing infrastructure support and the level of excess capacity which may exist will influence decisions concerning expansion of the infrastructure system and its cost. For example, the Office of Technology Assessment study (6) projects that the present and planned structure of the communities of the Piceance Basin can support an oil shale industry of approximately 133,000 barrels per day without assistance. It is growth associated with a 400,000 to 500,000 barrel per day industry which concerns us.

Impact Studies

There have been several studies which have attempted to examine the magnitude of the socioeconomic impact associated with the oil shale production industry with in the Piceance Basin. A number of these are site- and single-plant-specific, examining the impact associated with a particular facility on a particular site. We have been unable to find a comprehensive assessment which examines the impact of all potential oil shale and energy development within that region and focuses on all elements of the analysis; that is, from technology deployment through definition of infrastructure support cost. Nonetheless, we can provide information which represents the most recent estimate of the socioeconomic impact of oil shale production within Colorado.

The previously established target for oil shale production was 400,000 barrels per day in 1990. The construction labor force required during the period 1980-1990, normalized to a 1990 production capacity of 400,000 barrels per day, has been estimated as 54,336 labor years of effort.

The total estimated operations employment over the same period, again normalized at 400,000 barrels per day production, is 57,114 labor years of effort.

This increase in population due to in-migrating workers also stimulates other population and economic growth. There are a series of multipliers which are used to estimate this growth. These multipliers are related to:

	LOW CASE	HIGH CASE
EMPLOYMENT		
CONSTRUCTION	5,433	5,433
OPERATION	5,712	5,712
TOTAL	11,145	11,145
MULTIPLIERS		
FAMILY (F)	1.2	3.8
INDIRECT EMPLOYMENT (I)	0.4	2.0
POPULATION		
DIRECT EMPLOYMENT	11,145	11,145
DIRECT EMPLOYMENT FAMILY	13,374	42,351
INDIRECT EMPLOYMENT	4,458	22,290
INDIRECT EMPLOYMENT FAMILY	5,350	84,702
TOTAL	34,327	106,488
PER CAPITA		
INFRASTRUCTURE $(C)	3,00.00	6,000.00
TOTAL		
INFRASTRUCTURE $(S)*	103 MILLION	963 MILLION

$$*S = PC [(1 + F) (1 + I)]$$

Figure 4. Range of infrastructure costs for period 1980 through 1990.

- Worker average family size.
- Indirect employment generated.
- Infrastructure support cost.

A wide range of multipliers have been used to estimate family size associated with these new workers. The multipliers found within related studies are as low as 1.2 and as high as 3.8. It is important to note that it may be especially appropriate to segregate estimated family size by the construction and operations phase. Construction workers and their families are often transients and do not establish permanent residence within the geographic area experiencing development. The average number of dependents per construction worker has been estimated as 1.2 per worker; that for operational workers as 2.4 per employee.

Secondary or indirect employment generated by the introduced oil shale industry is also represented by a range of multipliers. These multipliers are as low as 0.4 and as high as 2.0. Once again, as in the case of estimating associated family size, segregation of these multipliers for construction and operations is appropriate; the range of multipliers that we have seen for both categories is coincidental with the range mentioned above.

Based upon the use of these data, the range of total 1990 population growth associated with oil shale production in Colorado has been estimated from a low of 54,719 to a high of approximately 85,382 for a normalized 400,000 barrel per day case. We have only seen one cost estimate for the related infrastructure support. This estimate was prepared by the Colorado State government, and represents the incremental per capital cost of infrastructure related to energy production. The cost estimate is $4,725 per person. This falls within a commonly used infrastructure support cost range which we have become familiar with of $3,000 to $6,000 per person that studies of other areas have estimated.

It may be demonstrative to use these multipliers in their extreme to calculate the range of infrastructure support cost which could result from their use. Figure 4 provides the particular mulipliers used and the result. The calculated range is $100 million to approximately $1 billion, an order of magnitude difference.

Future Efforts

The information presented here demonstrates that there certainly is a fairly significant range of specific

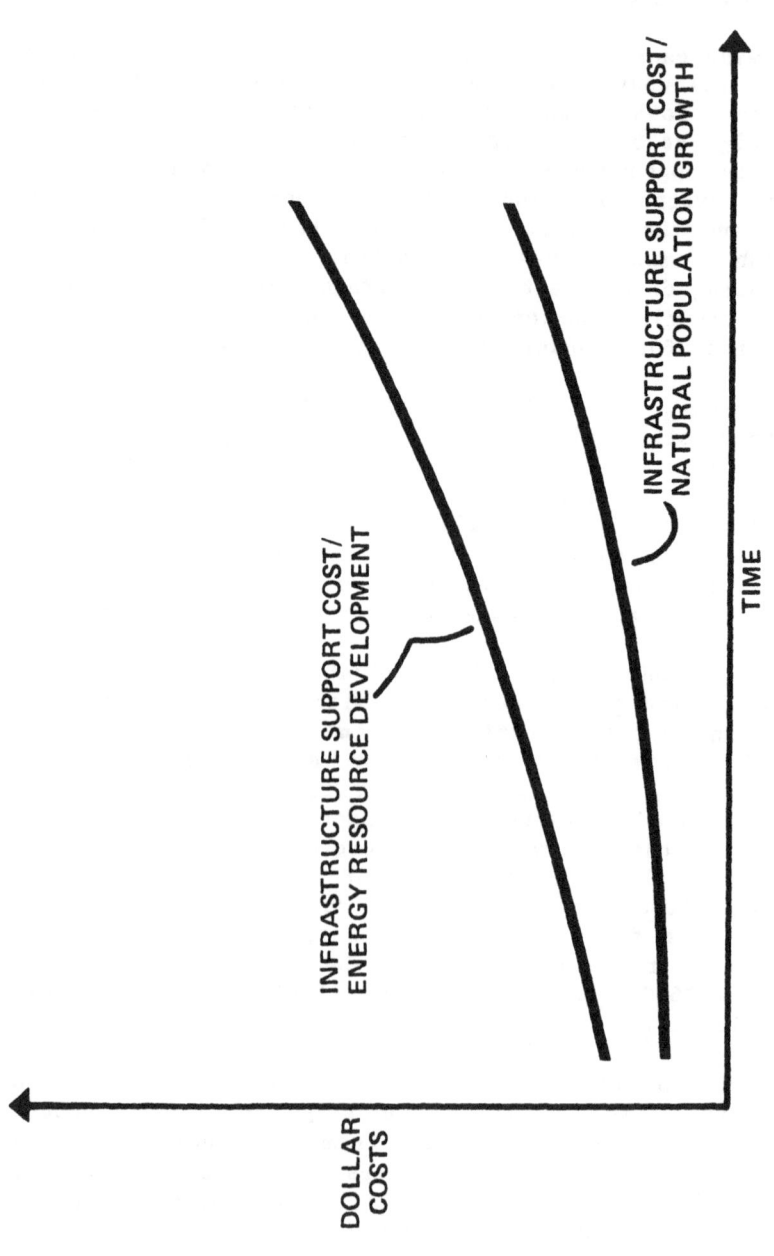

Figure 5. Infrastructure cost comparison

conclusions concerning socioeconomic impacts associated with oil shale production. Of course, as we move through the demonstration phase to commercialization for the actual facilities, we will learn more about the technologies and their associated impacts, mitigation plans can be refined.

Our expectation is that these uncertainty ranges will become smaller between now and 1990. There are several reasons why these data ranges are broad. First, the oil shale technologies are still in their embryonic stage, and we have yet to learn specific data about their construction needs and operating characteristics, including associated employment. The various multipliers mentioned above should become more specific factors as we learn more about the technologies and the geographic area's reaction. We feel that the growth pattern which will be experienced will not be similar to the "boom-bust" situation which often accompanies the introduction of a single plant within a rural area, but rather rapid growth maturing to continuing urbanization. As time passes, the exact patterns which we can use as a basis for forecasting will become evident.

We feel that there is a basic error in approach in attempts to estimate the required infrastructure support cost that inflates the estimates. But this discrepancy can be reduced by proper planning and management. The error lies in estimating the incremental infrastructure support cost resulting from energy resource production without significant proper attention to revenue collection patterns. The traditional method, represented by Figure 5, requires the comparison of the estimated infrastructure support cost for normal, non-energy related growth with the growth that would occur with an oil shale or other energy production industry introduced in the area. The incremental cost is calculated, divided by the incremental population growth, yielding a per capita cost of additional infrastructure support. It is our conclusion that a better approach would be to compare that incremental cost to the incremental revenue which would result from energy resource development.

This method would require the estimation of incremental infrastructure cost **and** incremental revenue flowing from the placement of an oil shale production industry within the geographic area. We can then make a comparison of cost versus revenue. Early in the growth of an area, infrastructure support costs will most likely outspace available revenue. But, as the area becomes more urban and stable in character, the relationships will change; revenue will outpace costs. Finally, as the area continues to grow, a diseconomy of scale may occur and cost will again (and possibly always) outpace revenue. This last stage is where most urban areas within our nation find themselves today. The middle stage poses no revenue problem. The appropriate concern, then is with the first stage of this relationship. The specific questions which require answers are: what is the magnitude of the cost-revenue differential? Over what period of time does it occur? How can compensation for the revenue short-fall be obtained? In short, this is the "up-front" revenue problem.

Problems of New Estimates

This paper has spent considerable time discussing the uncertainties surrounding the attempt to predict the impacts (especially environmental impacts) of major public and/or private decisions. Although the professional literature speaks in such knowing terms as "contingency planning," in reality, because the current art of impact analysis is so uncertain, the possibility of a change in the perturbation is downplayed or ignored. Today, there is less push for public involvement in the synfuels area and there is change in the MX missile plans. All of this will mean more deployment guesses for where growth may occur and more guesses as to how significant it may be. From the analyst's perspective, these shifts are likely to be annoying but they can be dealt with. It is hard to see, though, what the local politicians, planners, and administrators will do with such moving targets. If they truly believed the longer range projections and acted responsibly on this information, then they may very well have set in motion expensive projects which will have to be significantly revamped or even scrapped later on. If major policies change rapidly and impact a given area of the country heavily, those who must deal with the change have the choice of ignoring the analysis of the impact in the hope that they can deal with whatever issues arise on an incremental basis. Either way seems to be a problem; and, as with lots of other things in life, we need to develop a method so that everyone can be satisfied.

References

1. Los Alamos Scientific Laboratory and Abt/West. Draft document prepared for the Office of Environmental Assessments, U.S. DOE.

2. Peter W. House, Office of Environmental Assessments, U.S. DOE. Testimony before the Senate Budget Committee, Rifle, Colorado (August 29, 1980).

3. Anon., Synthetic Liquid Fuels Development: Assessment of Critical Factors. ERDA 76-129/2, Vol. II p. 579 (1976).

4. Anon. Environmental Analysis of Synthetic Liquid Fuels. DOE/EV, P. 62 (July 1979).

5. Anon. Synthetic Fuels and the Environment: An Environmental and Regulatory Impacts Analysis. DOE/EV, Ch. V (June 1980).

6. Office of Technology Assessment, U.S. Congress. An Assessment of Oil Shale Technologies. pp. 4, 256 (June 1980).

7. General Accounting Office. Water Supply Should Not Be an Obstacle to Meeting Energy Development Goals. (January 1980).

8. U.S. Department of the Interior. Water for Energy in the Upper Colorado River Basin. (1974)

9. Colorado Department of Natural Resources. Colorado Oil Shale: the Current Status. (1979)

10. Denver Research Institute. Predicted Costs of Environmental Controls for a Commerical Oil Shale Industry. (1979)

11. General Accounting Office, Report CED-80-30.

12. Colorado Department of Natural Resources. The Availability of Water for Oil Shale and Coal Gasification Development in the Upper Colorado River Basin. Draft Report. (1980).

13. Univ. of Oklahoma. Energy in the West. Report prepared for U.S. EPA (1979).

Part 4

The Social and Economic Impacts of Western Energy Development

William R. Freudenburg

12. An Overview of the Social Science Research

Introduction

The Rocky Mountain and Northern Great Plains states of the western U.S. are currently facing some of the most important and far-reaching changes in the history of the region. As should be clear from other sections of this volume, large-scale resource developments have the potential to create significant impacts on the environment and natural resources of the intermountain West; as the chapters in this section of the volume will indicate, those same developments have important implications for the region's people, as well.

Until quite recently, the social or "people" impacts—including economic, political, psychological and cultural impacts as well as the strictly "social" ones—have received far less scientific scrutiny than have physical or biological impacts. The problem is not that the social impacts cannot be studied scientifically; it is instead that they have been guessed at more often than they have been studied.

Perhaps this has occurred in part because virtually everyone feels the right to claim a certain degree of expertise about "human nature," while fewer are quick to make pronouncements about Mother Nature. Perhaps it reflects the fact that social impacts are of concern to the broader public

*The author thanks Linda M. Bacigalupi and Pamela D. Savatsky for helpful suggestions on an earlier draft of this chapter. This is Scientific Paper No. 6553, Research Project 0478, Agricultural Research Center, Washington State University. Portions of the discussion are drawn or developed from some of the related papers of the same Research Project (see especially Freudenburg, 1982b, 1982c).

even when they are not being studied--meaning that assertions flow in to fill the gaps that exist when the necessary information is unavailable. Whatever the reasons, it is possible to find a broad variety of allegations about the likely social consequences of developments. Some commentators claim that energy development will be a dream come true for rural areas of the West; others claim that the developments will be the worst kind of nightmares. Evidence for either kind of assertion is notable mainly by its absence.

Within the last decade, however, the ratio of evidence to assertion has begun to improve. Several factors have helped to increase social science research in this area in recent years, but perhaps the most obvious factor is a matter of law--specifically, the "National Environmental Protection Act of 1969," or NEPA (P.L. 91-190, 42 U.S.C. 4321-4347).

This is the law that led to the requirement for environmental impact statements (EISs). It is also a law that calls for an increased use of social science expertise in environmental decision making. Section 102 of the act requires federal agencies to make "integrated use of the natural *and social* sciences" in "decisionmaking which may have an impact on man's environment" (emphasis added). Section l02 (C) requires that EISs--including "integrated use" of the social sciences--be prepared for all "major federal actions significantly affecting the quality of the human environment." The President's Council on Environmental Quality, which has the responsibility for guiding the implementation of the law, has further clarified legal intent. Section 1508.8 of the regulations for implementing NEPA explicitly notes the need to consider social as well as ecological and other impacts, "whether direct, indirect or cumulative" (U.S. Council on Environmental Quality, 1978:28; see also the discussions in Friesema and Culhane, 1976; Meidinger and Schnaiberg, 1980; Freudenburg and Keating, 1982).

Other factors have also helped encourage increased social science research in this area, as will be noted in the next chapter, by Stan Albrecht. For the present, however, it is sufficient to note the net result--the emergence of an important and growing body of social science literature on the impacts of western energy (and other resource) developments. This body of work is far from complete, to be sure, yet it is already sufficiently well-developed that it could not be fully summarized in the limited space available in this volume.

Nor is that our purpose. Rather than trying to summarize everything that is now known about the social impacts of large-scale developments in the West, these several chapters are intended to provide a selective sampling--one that provides examples of the types of findings already accumulated and that represents the types of work presently underway. Before turning to the specific chapters, accordingly, it would be useful to provide a brief overview of that larger body of work.[1]

The Nature of "Socioeconomic Impacts"

One of the first problems of discussing the social or "socioeconomic impacts" of western developments is that the terminology can imply so much that it actually communicates very little. At one extreme, "socioeconomic impacts" include all of the changes caused by a development that are not explicitly physical or biological in nature. At the other extreme, "socioeconomic impacts" are a specific type of strictly economic impacts. In general, however, the term tends to be used to describe impacts upon three different aspects of the human environment--the demographic/economic environment, the social/cultural environment, and the biophysical environment. (The categories used here are modified from those apparently first suggested by Albrecht, 1978; see also Freudenburg, 1982b.)

1. *Demographic and Economic Impacts--changes in the provision of facilities and services.* This first category may be the most common meaning of the phrase, "socioeconomic impacts," and it is probably also the most common branch of "socioeconomic impact assessment." The variables involved here are usually demographic and economic in nature. The usual concern is that when a community's population grows suddenly and dramatically--a demographic change--there can be an equally sudden and dramatic increase in the demand (and economic cost) for the kinds of facilities and services normally provided by local governments--water and sewage treatment, streets, police and fire protection, and so forth.

2. *Socio-Cultural Impacts--changes in individual, group, family or community functioning.* At the same time that a facility is causing disruptions to the local service providers, however, it may also be disrupting the kinds of "services" that people normally provide for one another on a relatively informal basis--through effectively functioning families, neighborhoods, friendship groups, and so on. The signs of problems in this area can show up in increased rates of crime and juvenile delinquency, drug abuse, worker turnover, and other indicators of social disruption. The variables that may need to be analyzed to provide an adequate assessment of the changes may include intergroup and intragroup conflicts,

FIGURE 1.
THE MAJOR TYPES OF LOCAL "SOCIOECONOMIC" IMPACTS:
EXAMPLES OF QUESTIONS ASKED

Aspect of the Human Environment	IMPACTS	EVALUATIONS
Demographic/ Economic	* How many new people will a project bring to the area? * Do local facilities and services need to be expanded or upgraded? * Will local unemployment rates decline or increase?	* Do present citizens want or object to growth? * What level of service is considered "adequate" by the incoming population? * Will local merchants object to paying higher wages to attract employees?
Social/ Cultural	* Will conflicts erupt between oldtimers and newcomers? * How much stress will be created for families and young persons? * Will there be increases in crime and delinquency, alcoholism, and other social problems?	* Will locals feel that the comfortable character of their town has been lost? * Will young people want to stay in a larger community? * Are most locals willing to accept increases in crime and other problems in exchange for increasing economic vitality?
Biophysical	* How many ranching families will be displaced by strip mines (or new housing developments?) * Will the projects' resource use or environmental impacts preclude other developments in the future? * Will opportunities for solitude decline?	* Will local residents or future tourists find strip mines to be highly objectionable esthetically? * Are locals willing to institute the kinds of laws and regulations that would effectively limit environmental degradation? * Will people mind if opportunities for solitude decline?

community-level interaction patterns, personal stresses, and a broad range of human behaviors that have very little direct connection to the number of dollars in a person's pocket.

3. *Biophysical impacts--The human consequences of projects' direct and indirect impacts on the biophysical environment.* This may be the least well-defined area of socioeconomic impact assessment, perhaps partly because some of these impacts are essentially biological or epidemiological (e.g., the health hazards created by air pollution from power plants), while other involve some mixture of social and physical forces (e.g., when ranch families lose their irrigation water to large new energy developments). Still others fall into areas that are not within the expertise of any existing discipline. The impacts in this category include the various ways in which humans are affected by changes in the biophysical environment--both the changes caused "directly" by the development in question and those caused "indirectly" by new housing developments or other changes that have themselves resulted from the development.

The categories of this three-way division are not perfectly and completely distinct from one another, and other approaches to such a typology are possible. The impacts in these three categories, however, do tend to be quite different from one another most of the time--and it is almost never the case that by dealing with one type of impact one has somehow "taken care of" the other two as well. While the three kinds of impacts tend to be interrelated, in short, they are anything but identical or interchangeable. Our discussion is thus likely to become quite vague if we allow the three to become confused.

As Figure I shows, moreover, social science research in all three categories (or "aspects of the human environment") can deal with two relatively distinct questions. The first has to do with "impacts"--the actual changes that result from a given development. The second has to do with what are here called "evaluations"--people's evaluations of and attitudes toward the impacts.

Persons who are not familiar with existing social science research often tend to assume that there are essentially two categories of socioeconomic impacts--economic impacts, which can be evaluated scientifically, and "lifestyle concerns," which are essentially just matters of taste, and which are thus not amenable to scientific analysis. Figure I is an attempt to illustrate that, by contrast, while the *impacts created* are often very different from people's *evaluations of* the impacts, it is not true that only the economic impacts can be evaluated dispassionately, or that only the social or the biophysical impacts have an evaluative component. Both the actual impacts and the evaluations of the impacts can be found in all three aspects of the human environment.

Two further points about Figure I are in order here. The first is that social scientists have no greater expertise in matters of esthetics than do any other observers. Social scientists--at least in their roles as social scientists--cannot say that community growth is "good," that the replacement of locally owned businesses by chain stores is "bad," or that a given form of land reclamation is beautiful or esthetically distasteful. What social scientists can do is to act on the observation that "beauty" (and other evaluations) will tend to be in the eye of the beholder. Established methods allow the evaluations of local residents (or other "beholders") to be reported in a reliable and accurate fashion, and thus to be considered, where appropriate, along with information on the social, economic and other impacts of a proposed development.

The second point is that--in recognition of the fact that social science research on biophysical impacts is not nearly as well-developed as research in the other three areas--the biophysical impacts will receive little further discussion in the present section of this volume. The omission is not intended to indicate that these biophysical impacts are unimportant--only the fact that they have not yet been studied as carefully as the impacts in the other two areas.[2] It is to be hoped that social science research on biophysical impacts will continue to develop, leading to a body of information that is too substantial to be omitted from future summaries such as this one. The chapters in this volume will focus instead on the two areas of research that have seen greater activity to date. One chapter will report on recent findings in the areas of demographic/economic impacts; one chapter will report on recent findings in the area of social/cultural impacts. Like most of the writings in the broader literature, both of these two chapters concentrate mainly on the region's White or Anglo communities. The third chapter, accordingly, will focus on the special kinds of impacts that can occur when development affects Indian or Native American communities. The next portion of the present chapter will provide a more detailed introduction to the three chapters that follow.

Demographic and Economic Impacts

Chapter 14, by Steve Murdock and his colleagues, looks at the demographic (population) changes brought about by large-scale developments, and then compares these impacts with previous population projections. The size and speed of the population growth are major driving forces behind many of the impacts discussed in this section; most often, however, the population changes are of concern because of their implications for an important category of economic impacts-- "fiscal impacts," or the economic impacts on local governments. In brief, problems in this area arise because the new people are likely to want certain things--including the kinds of facilities and services normally provided by local governments.

The complicating factor is the so-called "front-end" problem: Many of the new services--particularly those that are normally provided by municipalities, such as sewers, water systems and streets--require substantial capital outlays. The new people are likely to want the services in place by the time they arrive; yet they are often not going to pay enough in taxes to provide the necessary capital until several years have passed (if then). The net result is a kind of Catch-22 for local officials. The officials need money "at the front end" for building the new facilities on schedule, but they are not usually able to get the money through normal channels until several years later.[3]

The techniques for dealing with the "front-end" economic and demographic impacts are relatively sophisticated. There is a broad range of variation in the specific techniques (and results) available, but most of the commonly used models take the same general approach. They start with the developer's estimate of the number of "new" workers to be brought into the area by the facility in question. Next they use a series of "multipliers" to take account of the number of new workers who will bring spouses and children, and of the number of additional "service" (or "nonbasic" or "ancillary") workers who are likely to come to town to help meet the demands of the new workers and their families. Putting all of these numbers together can lead to some overall quantity that represents a best guess at the number of new people who will be brought into town (both "directly" and "indirectly") because of the new development. This estimate--if it is accurate-- allows researchers to compute the resultant "demand" for services, and to compare that demand with the community's existing capacity for sewer and water treatment, classroom space and other service provision. Chapter 15, by Dr. Murdock and his colleagues, assesses the accuracy of the population estimates that have been contained in environmental impact statements to date.[4]

Four more points need to be made about economic and demographic impacts in the context of this volume. The first is that they can be worsened by jurisdictional mismatches--by the fact that our political boundaries often do not have the courtesy to be in the "right" places. There are numerous cases where the resources lie in one county, but where a community needing to deal with the growth and problems lies in another. Occasionally, the income and the impacts can occur in different states. Jurisdictional mismatches frequently occur even when the resources are located in the "right" county, however, simply because the mineral resources and the energy facilities themselves tend to be located well outside of any city boundaries. While the facilities may provide a substantial tax benefit to the counties and school districts in which they are located, in other words, they are not likely to pay taxes directly to local municipalities. Yet many of the truly expensive and capital-intensive services (such as streets or water and sewage treatment) are traditionally provided by municipalities.

The second point is that the economic impacts are inherently mitigable. Problems usually occur when a local community does not have enough advance warning or enough dollars to be able to meet the increased demand; thus the problems can usually be dealt with if energy developers are straightforward and helpful, if the workforce and population projections are accurate, if the various actors in a local community are cooperative, and if the proper number of dollars are there when they are needed.

The third point is that recent economic impact mitigation efforts fully seem to deserve the favorable attention they have been receiving. Substantial progress has occurred in the span of only a few years. Several recent efforts have shown that it is possible for industry, government and researchers to work together cooperatively and constructively; experience with recent success stories has shown that the results of such cooperative efforts can be beneficial to all the parties involved (Metz, 1980).

The fourth and final point is that the economic and demographic projections appear to be the best-developed aspect of socioeconomic impact assessment--but that all of us would be well advised to treat any given projection with a certain degree of caution. The compiler of a major recent volume offers this summary: "After two days of papers and dialogue, the predominant messages of the conference seemed to be: 'While computer models may be necessary, they are definitely not sufficient for assessing the impacts of major infrastructure and facility investments on rural communities,' and 'Use models, but don't believe them'" (McDonald, 1980: viii). While the models are often impressive in their complexity, in short, their projections are not yet perfect. Active practitioners in the area have often warned that the apparent precision of population estimates could be deceptive, but the chapter by Dr. Murdock and his colleagues presents what may be the most solid empirical test of the projections yet made available. The results of their efforts show us that previous caution and even skepticism may have been warranted. The previous inaccuracies also lend increased importance to the standards these authors suggest for future impact assessments.

Social and Cultural Impacts

The demographic impacts summarized in Dr. Murdock's chapter can be seen as "impacts on populations"; as noted above, they are normally of concern because of the resultant impacts on governments. Chapter 14, by Stan Albrecht, looks at another broad category of impacts that can result from rapid growth--socio-cultural impacts, or "impacts on people."

As noted in Figure 1, the actual socio-cultural impacts are conceptually distinct from people's evaluation of the impacts. Unfortunately, when environmental impact statements have considered impacts on local populations in the past, the documents have often focused primarily or exclusively on questions of attitudes--asking whether or not people *liked* the developments, rather than *how they were affected by* the developments. Albrecht's chapter, like a good deal of recent research, deals with impacts that go beyond what might be called "impacts on attitudes" and considers "impacts on lives." The emphasis is appropriate; empirical impacts on people's lives have often been ignored and have often been exaggerated in the past, and it is only through increased documentation that we will be able to obtain a more realistic assessment of the changes that actually do or do not take place. (For a recent debate on the data in this area, see Wilkinson et al., 1982, and the responses to it--e.g., Finsterbusch, 1982; Gale, 1982; Murdock and Leistritz, 1982).

Persons outside the social sciences sometimes have difficulty seeing the difference between changes that are merely "matters of esthetics" (or of evaluations) and the changes that "affect people's lives." Indeed, the dividing line can often be indistinct.[5] As a relatively straightforward rule of thumb, however, we might say that an "esthetic" impact becomes a matter of "significant social change" if it in some way impairs a person's ability to function.

Some of the statistics from boomtowns indicate that the degree of impairment may sometimes be quite substantial. For example, while Gillette, Wyoming, may have received more than its fair share of notoriety, the statistics from that community still deserve our attention. Kohrs (1974) reported that the community's suicide-attempt rate rose to the point where there was one attempt for every 250 persons; a later government document, which took pains to choose words less inflammatory than those utilized by Kohrs, noted that as many as 12% of the entire county's population during a boom period--approximately one person out of every eight--"(could) be characterized as having developed a 'drinking problem'" (U.S. Department of Interior et al., 1974:1.434).

Later evidence indicates, moreover, that the problem might not be any less severe "the second time around." The Northern Wyoming Mental Health Center did a survey of Gillette during a later period of boom growth in 1977, using the Holmes and Rahe social readjustment rating scale (Holmes and Rahe, 1967), and discovered that their community sample had a *mean* or average score of 308 "life change units" (Pattinson et al., 1979; Weisz, 1979). That average falls into the category found by Holmes and Masuda (1974) to indicate a "major life crisis" and a significantly increased likelihood of physical illness. In the Holmes and Masuda study, 49% of the persons with scores over 300 experienced illness in the nine months after the survey; only 9% of the persons with scores under 200 reported any illnesses over the same time period (Holmes and Masuda, 1974:64). The mental health center caseload in Gillette's county (Campbell) grew by slightly over 100% (from 397 to 799) in the three years from 1975 to 1978, compared to a 62.5% increase in county population, and to an average increase of only 11% in the mental health caseload of the four other counties in the Northern Wyoming region. Campbell County's admissions to the state mental health hospital (generally the more severe cases) grew 610%, from 10 to 71, between 1974 and 1978, while admissions from other area counties increased by 70.5% during the same period (Weisz, 1979: 36-37, 41-42).

In Rock Springs/Sweetwater County, Wyoming, where an energy boom led to a doubling of population between 1970 and 1974 (from 18,391 to 36,900), the local mental health center caseload increased approximately nine times. In the meantime, productivity in some established operations dropped 25% to 40%, as annual employee turnover climbed past the 100% mark for at least one of them. Cost overruns on the new plant that was one major source of the influx ran past 33%, and complaints to a local law enforcement agency increased by 60% in a single year; the overall number of crimes rose by more than 300% between 1969 and 1974 (Gilmore and Duff, 1975:2, 12-15; Lovejoy, 1977:15; Edgley, 1979).

In Craig, Colorado, where the construction of a coal-fired power plant also caused a population increase of about 100% (roughly from 5,000 to 10,000 persons) between 1974 and 1978, family disturbances rose by 352%, from 27 to 122, and child behavior problems rose an even 1000%, from 6 to 66; alcohol and other substance-abuse cases increased by 623%, from 13 to 94. The overall number of crimes per year rose 300%-450%, depending on the method of computation employed; or a two-month basis, crimes against property rose 222%, from 59 cases to 187, and crimes against persons were up 900%, from 4 to 40 (McKeown and Lantz, 1977; Lantz and McKeown, 1979; Freudenburg, 1979).

Spectacular increases in percentages generally need to be interpreted with caution, particularly if the "base" for computations is quite small (Freudenburg, 1982a); as in the case of the population projections, in other words, these numbers can be considerably less precise than they seem. Some of the greatest percentage increases are based on very small numbers of pre-boom cases; for example, the 1000% increase in child behavior problems in Craig, Colorado, stands for an increase from 6 to 66 cases. The increase seems to be quite real (and most of the other increases are based on a larger number of pre-boom cases), but a fluctuation of just a few cases in the pre-boom caseload could have led to substantial variations in some of the exact percentage increases computed here.

Other factors can also distort these figures; most notably, agencies sometimes change accounting procedures in a way that exaggerates (or masks) the changes taking place, and given individuals can sometimes be "double-counted," showing up as "alcohol abuse" as well as "property crime" cases, for example. In short, most social scientists who are familiar with the "dismal list" of boomtown statistics will warn that the individual statistics need to be treated with a good deal of caution; confidence is gained largely from the broader pattern of findings, rather than from any single figure.

The broader pattern of findings, however, reads almost like a classic list of "social pathology" indicators, and therein lies another potential danger. A number of writers have evidently been so carried away by the pattern of findings that they have waxed eloquent about presumed "boomtown pathologies," leaping to conclusions that go well beyond the pattern of documentation that has accumulated--asserting for example that the boomtown dweller "lives under conditions worse than those found in an urban ghetto" or that boom growth is a form of "social and cultural genocide." Such assertions may help headline writers find colorful quotes, but they are as unrealistic as the equally naive assertion that all is well in the energy boomtown.[6]

The issue of "social pathologies" is sufficiently important that it is treated in some detail in Chapter 14. Dr. Albrecht has deliberately shortened and simplified this discussion to keep it appropriate for an audience of nonspecialists; Albrecht's conclusions, however, would appear to be shared by the majority of the persons who are actually familiar with the research and the theoretical tradition in question. Classical social theories on social disruption do have a certain degree of relevance to understanding rapidly growing communities, but only if those theories are used cautiously. They are more useful in sensitizing us to remember to consider certain variables than in "proving" that certain outcomes are likely; in the social sciences, as in any other form of science, documentation of outcomes is after all what research and evidence are meant to provide (see also Cortese and Jones, 1977, 1979; Cortese, 1980; Little and Lovejoy, 1977; Moen et al., 1981; Freudenburg, 1980). After considerable examination of the evidence, Albrecht concludes that the social disruptions of rapid change should not be exaggerated, but that the changes "are very real." The preponderance of existing quantitative evidence, in my own view, clearly supports his conclusion. His research also leads him to suggest that one factor behind the impacts is the simple (but often misunderstood) variable of a person's length of residence in a community.[7]

The Special Cases of Native Americans

The paradoxes of western energy developments are in evidence throughout this volume, but perhaps nowhere are they more visible than when the developments affect Native Americans, or Indians. In addition to the types of impacts often seen in White communities, the Native Americans can often experience impacts that are very different in nature, being shaped by the relationships between the reservations and the rest of the nation, by the relationships among the Native Americans themselves, and by the often-ignored but often significant cultural differences that exist. Chapter 16, by Joseph Jorgensen, summarizes the growing literature in this area.

As Dr. Jorgensen's chapter points out, reservations in the Western U.S. have many deposits of energy resources—including coal, oil, natural gas, uranium and so on—which are far too important to be ignored. The chapter also points out that Native American cultures may also have a human importance which ought not be ignored, but which has evidently been overlooked on many occasions in the past. The consequences of such failures are often quite negative, as Jorgensen's chapter shows.

The Native Americans tend to be among the most disadvantaged of all Americans, generally having incomes that are well below the poverty level and often living under conditions that seem little short of desperate by contemporary standards. Given the tremendous mineral wealth that lies below many of the reservations, it is scarcely surprising that many commentators have viewed resource development as providing the key to an improved quality of life for the reservations' inhabitants. What may be surprising to many readers is Jorgensen's assessment of the consequences of past efforts to develop the resources.

While Native Americans often need the economic advantages promised by development, they generally stress that they do not want the development to occur at the expense of environmental quality and their tribal heritage. Dr. Jorgensen's assessment, however, is that Native Americans have generally realized few of the economic benefits of development, while often suffering negative consequences that have been far greater than they were originally predicted to be. His assessment is critical, but it is extensively supported through references and examples, and it reflects his thorough familiarity with the research conducted to date. His evidence, in short, cannot simply be ignored. His chapter reminds us that business as usual--including apparently "rational" business decisions that are not deliberately intended to harm anyone--can sometimes have important unforeseen consequences. His chapter also provides us with an unmistakable challenge to do better in the future.

Summary and Conclusions

It is useful to see the socioeconomic consequences of western developments as falling into three relatively distinct categories of impacts upon the human environment: demographic and economic impacts, socio-cultural impacts and biophysical impacts. There are differences as well as similarities in the types of impacts that are likely to be created in each of the three areas.

Overall, given the findings that are now available from previous research, developments of the scale currently envisioned for the West appear likely to lead to a broad range of significant and occasionally even severe impacts. Both in the area of demographic and economic impacts and in the area of social and cultural impacts, we can start to outline the general nature and magnitude of the impacts with a reasonable degree of confidence. The social aspects of developments' biophysical impacts cannot be projected with the same degree of confidence at present, but these impacts often have the potential to be substantial as well.

Many of the impacts will probably be lessened or even overcome by the mitigating measures now being developed; other impacts, however, are not likely to be mitigated. First, in the area of demographic and economic impacts, it now appears that many of the negative impacts of growth can be mitigated with a reasonable level of success; indeed, the efforts that have already gone into economic impact mitigation are genuinely noteworthy, showing substantial progress in a very few years. Second, in the area of biophysical impacts, further research will be needed before the appropriateness of available mitigating strategies can be determined. Finally, in the area of social and cultural impacts, it appears that only some of the likely disruptions are being addressed, even in the more impressive "state of the art" mitigation programs. Part of the reason for the frequent oversight, it appears, is that the more obvious economic impacts are normally the first to be addressed; another part of the reason may be that many of the socio-cultural impacts of energy development are a relatively direct function of the speed and scale of development. In short, some of the most significant socio-cultural impacts may be essentially unmitigable if energy developments do indeed proceed along the lines currently envisioned. Moreover, as Cortese and Jones (1977) have pointed out, some of our very efforts to solve economic impacts ("more and better planning," increased professionalism and bureaucratization, and so on) may actually contribute to the severity of some socio-cultural impacts--the increasing bureaucratization, professionalization and "depersonalization" taking place.

Unanswered Questions

The data in these chapters raise two questions that are particularly important in the context of this volume; one of them is political and the other is scientific in nature. The political question is more obvious. The best available evidence indicates that the envisioned level of development will create significant disruptions for local residents, with many of the socio-cultural impacts probably not being mitigable at the proposed level of development. Given this evidence, have we as a nation actually decided that we wish to encourage "full-scale" western developments?

It is not the place of this volume to answer this question; by definition, the issue is a political rather than a scientific one. It is important, however, that the question be raised and addressed. It does not appear that the question has been fully considered at present; indeed, it appears that many persons (particularly in the eastern U.S.) may not yet have comprehended the full range and magnitude of the impacts that are likely.

The second question is less obvious, but perhaps equally important. These chapters are attempts to respond to a series of questions about the likely impacts of large-scale developments in the West. If those questions are asked again in the future--five or even ten years from now--will we be able to respond with better answers?

Current research efforts such as the one summarized by Dr. Murdock and his colleagues do hold the promise of providing improved information in the area of demographic and economic impacts (see also Gilmore et al., 1982). It appears that the necessary research on biophysical and on socio-cultural impacts, however, is not currently underway. The problem is clearly *not* one of the "scientific impossibility" of the task; particularly in the area of socio-cultural impacts, reasonably well-developed social science techniques do exist for providing improved documentation. The research must be performed and supported, however, before the documentation can be provided, and not even the more extensive monitoring studies have yet included provisions for collecting the types of information that will be needed. In short, while questions about socio-cultural impacts are not unanswerable, they cannot be answered in the absence of appropriate documentation. It is particularly worth noting that many of them cannot be answered on the basis of commonly available or Census-type information, just as questions about oil-rich geological formations, for example, cannot be answered with any "available data" on the average current price of a gallon of gasoline. Further research is required, and it is not currently being performed.

Most of us would find it inconceivable that engineers could be asked only to do "projective studies" (in designing bridges, for example) without ever checking later to find out whether or not their bridges actually stood up. Nor would we generally expect them to try to design bridges without having some very basic information--on the tensile strength of steel, perhaps, or on the effects of wind and weather, on the types of construction required to support different kinds of traffic, and so forth. In fact, the very idea of doing without this information seems a bit on the ridiculous side--and yet something very much like that has been happening with many of the project-specific impact assessments that have been done to date. Impact researchers are essentially asked to go into a situation and come up with an educated "best guess" about the types of impacts likely to be created by a given development--yet to do so without having much more to go on than the accumulated "best guesses" of those who have gone before. Most such studies, in turn, produce little more than one more guess.

In fact, despite the millions of dollars that have already been spent on "socioeconomic assessments," very few dollars have actually been invested in empirical research. The majority of the evidence summarized in this section comes from a relative handful of studies, some of which have been conducted at researchers' own expense; were it not for these few studies, we would have nothing to go on but guesswork and assertions. The needs of decisionmakers--as well as the canons of science and the requirements of longer-term rationality--would all seem to suggest that a different approach would be more fruitful for the future.

The cost of filling the significant research gaps would not be negligible, but it would also not appear to be excessive in view of the importance of the questions involved. Even an extremely high quality study could be done for well under 1% of the cost of most of the proposed developments, and a very good study could probably be done for less than one tenth of a single percent of the overall cost of most large-scale western developments.

The existing data base allows for somewhat more confidence than is sometimes realized, but it clearly still includes important weaknesses and gaps. These weaknesses in the data have caused many statements in these chapters to be far more tentative than might otherwise have been desirable. Even more undesirable is the fact that the existing data base has apparently encouraged some commentators on both sides of the development debate to make exaggerated claims that are anything but tentative.

Today's developments give us the opportunity to decide whether future development decisions will be based on evidence and analysis, or on allegations and exaggerations. If the necessary research is done, we will be able to respond to future questions with data; if not, we will continue to run the risk that decisions will be based on ill-informed assertions about "overwhelming" or "negligible" levels of impacts, and that the needed information will simply not be available. The social science chapters in this volume provide a reflection of the evidence that is in existence today; it is important for practical as well as scientific reasons that researchers be allowed to further improve our knowledge base in the future.

Footnotes

1. Entire books have been written to summarize recent research on the social and economic impacts of large-scale western developments; perhaps the best and most up to date of them is Weber and Howell (1982). Entire books have also been written on more specialized topics--for example, various techniques for projecting economic and demographic impacts (Leistritz and Murdock, 1981) or the delivery of human services in "energy boomtowns" (Davenport and Davenport, 1979). Interested readers are urged to consult these works, or others that will be cited in the chapters that follow, for a more complete overview of existing research.

2. One of the reasons for drawing attention to the biophysical impacts, in fact, is to take note of their present and potential importance. Another is to draw attention to the need to keep them conceptually distinct from the other two categories of impacts. For an overview of previous sociological work on relationships between humans and the physical environment, see Dunlap and Catton (1979a, 1979b). For useful discussions of esthetics, see Bagley et al. (1973); Tuan (1974); Zube et al. (1975).

3. A related problem may be at least equally significant, but it has received much less attention to date--the potential for "back-end" problems, both in the short-run and the long-run. A community can be faced with a short-run "back-end" problem if the construction workforce for the facilities is significantly larger than the "permanent" or operating workforce, or if the developer decides to cancel the project after construction is already underway. If the community builds enough facilities to meet the short-term demand, it may be left with substantial excess capacity later on. The second kind of "back-end" problem is less immediate, but it may have at least equal importance in the long-run; communities may also need to face the question of what they will do to support their accustomed level of development when a finite environmental resource eventually runs out, or when the energy facilities simply close, as many have done in Appalachia.

4. Gilmore et al. (1982) provide another important recent assessment. A number of sources discuss the "how to" and "why" of such assessments; examples include American Statistical Association (1977); Bronder et al. (1977); Bolt et al. (1976); Booz, Allen and Hamilton (1974); Cluett et al.(1977); Edgley (1979); Energy Impact Assistance Steering Group (1978); Fitzsimmons et al. (1977); Gilmore (1980); Gilmore and Duff (1975); Leistritz and Murdock (1981); Luther (1979);

McDonald (1980); Mountain West Research (1979); Murdock and Leistritz (1979); Murphy/Williams Urban Planning and Housing Consultants (1978); Office of Technology Assessment (1980); Payne and Welch (1981); U.S. Department of Housing and Urban Development (1976); U.S. Energy Research and Development Administration (1977).

5.   One of the problems is that we are often quite willing to draw attention to other people's "tastes" without being fully aware of our own.   Many of the near-sacred beliefs of our culture--the privacy of home and property, for example, or our constitutional privileges--might be seen as "mere matters of U.S. tastes" by persons from other cultures.   Yet social changes which disrupt such "tastes" could be profoundly dis-orienting to people in the United States, even though they might seem scarcely worthy of note to persons accustomed to different cultural systems. (Chapter 16, by Dr. Jorgenson will discuss this point at somewhat greater length.)

6.   The dramatic assertions are also probably counterproduc-tive, even for well-meaning persons who wish to call public attention to problems that do in fact exist.   Some commenta-tors have apparently concluded that since some "reports" of boomtown problems are exaggerated (or simply fabricated), even the legitimate social science reports on the impacts of rapid growth should simply be disregarded.   Such over-reactions, obviously, contribute no more to our understanding than do the overblown assertions of chaos.   What is needed in the area of socio-cultural impacts, as in the area of demo-graphic and economic impacts, is careful documentation and clear understanding of the impacts that do and do not occur.

7.   Again here, there is a good deal of evidence to support Dr. Albrecht's conclusion.   Some of my earlier work has drawn attention to a related but distinct variable--a communi-ty's "density of acquaintanceship," which is essentially a mea-sure of the proportion of people in a community who are acquainted with one another.   The rapid population influx in a western energy boomtown appears to lead to a significant decline in the community's density of acquaintanceship, although this does not imply that any given person will be cut off from new or existing friendships.   What it does seem to imply is that some of the community's informal "mech-anisms" or arrangements that depend on a high density of acquaintanceship (as in the case of controlling deviance in-formally, aided by the fact that "everybody knows everybody else") can be limited in the effectiveness of their functioning. For a more detailed discussion, see Freudenburg (1980).

References

Albrecht, Stan L.
1978   "Social Impacts of Energy Development in the West." Paper presented to annual meeting of American Association for the Advancement of Science, Washington, D.C., February.

American Statistical Association
1977   *Report of the Conference on Economic and Demographic Methods for Projecting Population.* Washington, D.C.: American Statistical Association.

Bronder, Leonard D., Nancy Carlisle and Michael D. Savage, Jr.
1977   *Financial Strategies for Alleviation of Socioeconomic Impacts in Seven Western States.* Denver, CO: Western Governors' Regional Energy Policy Office.

Bolt, Ross M., Dan Luna and Lynda A. Watkins
1976   *Boom Town Financing Study, Volume I: Financial Impacts of Energy Development in Colorado--Analysis and Recommendations.* Denver, CO: Colorado Department of Local Affairs.

Bagley, Marilyn D., Cynthia A. Kroll and Kristin Clark
1973   *Aesthetics in Environmental Planning.* Prepared for Office of Research and Development, U.S. Environmental Protection Agency. Washington, D.C.: U.S. Government Printing Office.

Booz, Allen and Hamilton, Inc.
1974   *A Procedures Manual for Assessing the Socioeconomic Impact of the Construction and Operation of Coal Utilization Facilities in the Old West Region.* Washington, D.C.: Old West Regional Commission.

Cluett, Christopher, Michael T. Mertaugh and Michael Micklin
1977   *A Demographic Model for Assessing the Socioeconomic Impacts of Large-Scale Industrial Development Projects.* Seattle, WA: Battelle Human Affairs Research Centers.

Cortese, Charles F.
   1982   "The Impacts of Rapid Growth on Local Organizations and Community Services." Pp. 115-135 in Bruce A. Weber and Robert E. Howell (eds.) *Coping With Rapid Growth in Rural Communities*. Boulder, CO: Westview.

Cortese, Charles F. and Bernie Jones
   1977   "The Sociological Analysis of Boom Towns." *Western Sociological Review* 8 (#1, August):76-90.

   1979   "Energy Boomtowns: A Social Impact Model and Annotated Bibliography." Pp. 101-163 in Charles T. Unseld, Denton E. Morrison, David L. Sills and Charles P. Wolf (eds.) *Sociopolitical Effects of Energy Use and Policy: Supporting Paper 5, Study of Nuclear and Alternative Energy Systems*. Washington, D.C.: National Academy of Sciences.

Davenport, Judith A. and Joseph Davenport III
   1979   *Boom Towns and Human Services*. Laramie, WY: Department of Social Work, University of Wyoming.

Dunlap, Riley E. and William R. Catton, Jr.
   1979a   "Environmental Sociology." *Annual Review of Sociology* 5:243-273.

   1979b   "Environmental Sociology: A Framework for Analysis." Pp. 57-85 in T. O'Riordan and R. C. d'Arge (eds.) *Progress in Resource Management and Environmental Planning*, Vol. I. Chichester, England: Wiley.

Edgley, Gerald J.
   1979   *Analyses of Current Laws, Methods and Measures to Regulate, Predict and Mitigate Socioeconomic Impacts Associated with Large-Scale Energy Development*. Unpublished Master's Thesis. Washington, D.C.: Department of Urban and Regional Planning, George Washington University.

Energy Impact Assistance Steering Group
   1978   *Report to the President: Energy Impact Assistance*. Washington, D.C.: U.S. Department of Energy.

Finsterbusch, Kurt
    1982   "Boomtown Disruption Thesis: Assessment of Current Status." *Pacific Sociological Review* 25 (#3, July):307-322.

Fitzsimmons, Stephen J., Lorrie I. Stuart and Peter C. Wolff
    1977   *Social Assessment Manual: A Guide to the Preparation of the Social Well-Being Account for Planning Water Resource Projects.* Boulder, CO: Westview.

Freudenburg, William R.
    1979   *People in the Impact Zone: The Human and Social Consequences of Energy Boomtown Growth in Four Western Colorado Communities.* Unpublished Ph.D. Dissertation. New Haven, CT: Department of Sociology, Yale University.

    1980   "The Density of Acquaintanceship: Social Structure and Social Impacts in a Rocky Mountain Energy Boomtown." Paper presented to annual meetings of the American Sociological Association, New York, August.

    1982a   "The Impacts of Rapid Growth on the Social and Personal Well-Being of Local Community Residents." Pp. 137-170 in Bruce A. Weber and Robert E. Howell (eds.) *Coping with Rapid Growth in Rural Communities.* Boulder, CO: Westview.

    1982b   "Theoretical Developments in Social and Economic Impact Assessment." Pp. 8-18 in Sally Yarie (ed.) *Alaska Symposium on the Social, Economic, and Cultural Impacts of Natural Resource Development.* Fairbanks, AK: University of Alaska.

    1982c   "The Social and Economic Impacts of Oil Shale Development: Research to Support Decisions." Pp. 272-295 in K. K. Petersen (ed.) *Oil Shale: The Environmental Challenges II.* Golden, CO: Colorado School of Mines Press.

Freudenburg, William R. and Kenneth M. Keating
    1982   "Increasing the Impact of Sociology on Social Impact Assessment: Toward Ending the Inattention." *The American Sociologist* 17 (#2, May):71-80.

Friesema, H. Paul and Paul J. Culhane
1976   "Social Impacts, Politics, and the Environmental
Impact Statement Process." *Natural Resources
Journal* 16 (April):339-356.

Gale, Richard P.
1982   "Commentary." *Pacific Sociological Review* 25
(#3, July):339-348.

Gilmore, John S.
1980   "Socioeconomic Impact Management: Are Impact
Assessments Good Enough to Help?" Pp. 278-
298 in J. M. McDonald (ed.) *Computer Models
and Forecasting Socio-Economic Impacts of
Growth and Development*. Edmonton, Alberta:
The Faculty of Extension, University of Alberta.

Gilmore, John S. and Mary K. Duff
1975   *Boom Town Growth Management: A Case Study
of Rock Springs/Green River, Wyoming*. Boulder, CO: Westview.

Gilmore, John S., Diane Hammond, Keith D. Moore, Joel J.
Johnson and Dean C. Coddington
1982   *Socioeconomic Impacts of Power Plants*   Palo
Alto, CA: Electric Power Research Institute
Report EA-2228, Research Project 1226-4.

Holmes, Thomas H. and Minoru Masuda
1974   "Life Change and Illness Susceptibility." Pp.
45-72 in Barbara S. Dohrenwend and Bruce P.
Dohrenwend. *Stressful Life Events: Their Nature and Effects*. New York: Wiley.

Holmes, Thomas H. and Richard H. Rahe
1967   "The Social Readjustment Rating Scale." *Journal of Psychosomatic Research* 11:213-218.

Kohrs, ElDean V.
1974   "Social Consequences of Boom Town Growth in
Wyoming." Presented to regional meetings of
the American Association for the Advancement
of Science, Laramie, WY.

Lantz, Alma E. and Robert L. McKeown
1979   "Social/Psychological Problems of Women and
their Families Associated with Rapid Growth."
Pp. 42-54 in U. S. Commission on Civil Rights

(ed.) *Energy Resource Development: Implications for Women and Minorities in the Intermountain West.* Washington, D.C.: U.S. Government Printing Office.

Leistritz, F. Larry and Steven H. Murdock
1981 *The Socioeconomic Impact of Resource Development: Methods for Assessment.* Boulder, Co: Westview.

Little, Ronald L. and Stephen B. Lovejoy
1977 *Western Energy Development as a Type of Rural Industrialization: A Partially Annotated Bibliography.* Monticello, IL: Council of Planning Librarians, Exchange Bibliography No. 1298.

Lovejoy, Stephen B.
1977 *Local Perceptions of Energy Development: The Case of the Kaiparowits Plateau.* Los Angeles: Lake Powell Research Project, Bulletin Number 62.

Luther, Joseph
1979 *The Price of Change: A Growth Assessment Handbook.* Cheney, WA: Partnership for Rural Improvement and the Department of Urban and Regional Planning, Eastern Washington University.

McDonald, J. Merril (ed.)
1980 *Computer Models and Forecasting Socio-Economic Impacts of Growth and Development.* Edmonton, Alberta: The Faculty of Extension, University of Alberta.

McKeown, Robert L. and Alma Lantz
1977 *Rapid Growth and the Impact on Quality of Life in Rural Communities: A Case Study.* Glenwood Springs, CO: Colorado West Regional Mental Health Center.

Meidinger, Errol and Allan Schnaiberg
1980 "Social Impact Assessment as Evaluation Research: Claimants and Claims." *Evaluation Review* 4 (#4, August):507-535.

Metz, William C.
1980 "The Mitigation of Socio-Economic Impacts by Electric Utilities." *Public Utilities Fortnightly* 106 (#6, September 11):34-42.

Moen, Elizabeth, Elise Boulding, Jane Lillydahl and Risa Palm
  1981   *Women and the Social Costs of Economic Development: Two Colorado Case Studies*   Boulder, CO: Westview.

Mountain West Research, Inc.
  1979   *A Guide to Methods for Impact Assessment of Western Coal/Energy Developments.*   Omaha, NE: Missouri River Basin Commission, Western Coal Planning Assistance Project.

Murdock, Steve H. and F. Larry Leistritz
  1979   *Energy Development in the Western United States: Impact on Rural Areas.*   New York: Praeger.

  1982   "Commentary." *Pacific Sociological Review* 25 (#3, July):357-366.

Murphy/Williams Urban Planning and Housing Consultants
  1978   *Socioeconomic Impact Assessment: A Methodology Applied to Synthetic Fuels.*   Prepared for the U.S. Department of Energy, Assistant Secretary for Resource Applications. Washington, D.C.:   U.S. Department of Energy.

Office of Technology Assessment, U.S. Congress
  1980   "Socioeconomic Aspects."   Pp. 419-473 of *An Assessment of Oil Shale Technologies*   Washington, D.C.:   Office of Technology Assessment.

Pattinson, Mick, Robert Weisz and Elizabeth Hickman
  1979   *Program Evaluation Reports: A Study of the Impacted Population, Related Stresses, Coping Styles, and Mental Health Program Needs in the Gillette, Wyoming Planning District.*   Sheridan, WY: Northern Wyoming Mental Health Center.

Payne, Barbara A. and Carol Welch
  1981   *An Assessment of the Socioeconomic Impacts of Energy Development in Northwestern Colorado.*   Argonne, IL: Argonne National Laboratory. (Draft Volume III of An Assessment of Energy-Related Development in the Western Oil Shale Region, Richard Olsen, Project Manager.)

Tuan, Yi-Fu
　　1974　*Topophilia: A Study of Environmental Perception, Attitudes, and Values*. Englewood Cliffs, NJ: Prentice-Hall.

U.S. Council on Environmental Quality
　　1978　*Regulations for Implementing the Procedural Provisions of the National Environmental Policy Act*. Washington, D.C.: U.S. Government Printing Office.

U.S. Department of Housing and Urban Development.
　　1976　*Rapid Growth from Energy Projects: Ideas for State and Local Action: A Program Guide*. Washington, D.C.: U.S. Department of Housing and Urban Development, Office of Community Planning and Development.

U.S. Department of Interior, U.S. Department of Agriculture, and Interstate Commerce Commission
　　1974　*Final Environmental Impact Statement of Proposed Coal Development in the Eastern Powder River Basin of Wyoming*. Washington, D.C.: U.S. Department of Interior.

U.S. Energy Research and Development Administration
　　1977　*Models and Methodologies for Assessing the Impact of Energy Development*. Washington, D.C.: U.S. Government Printing Office.

Weber, Bruce A. and Robert E. Howell (eds.)
　　1982　*Coping with Rapid Growth in Rural Communities*. Boulder, CO: Westview.

Weisz, Robert
　　1979　"Stress and Mental Health in a Boom Town." Pp. 31-47 in Judith A. Davenport and Joseph Davenport III (eds.) *Boom Towns and Human Services*. Laramie, WY: Department of Social Work, University of Wyoming.

Wilkinson, Kenneth P., James G. Thompson, Robert R. Reynolds, Jr. and Lawrence M. Ostresh
　　1982　"Local Social Disruption and Western Energy Development: A Critical Review." *Pacific Sociological Review* 25 (#3, July):275-296.

Zube, Ervin H., R. O. Brush and Julian G. Fabos (eds.)
　　1975　*Landscape Assessment: Value Perceptions and Resources*. Stroudsburg, PA: Dowden, Hutchinson and Ross.

13. Paradoxes of Western Energy Development: Socio-Cultural Factors

Introduction

During the 1970's, the combined forces of a rapidly increasing demand for energy and the desire on the part of the United States to achieve greater national self-sufficiency in meeting energy needs resulted in accelerating pressures to develop natural resources such as coal, oil shale, tar sands, and geothermal power sources. Since many of these important resources are located in what are basically rural areas of the Rocky Mountains and Northern Plains states, their development has had and will continue to have dramatic consequences for local communities in this region and for their inhabitants. This chapter will review some of these consequences.

It is important to recognize that energy development is not the only force that is affecting rural communities of the west. In fact, it is difficult to really understand the consequences of energy development without considering the broader context in which it is occurring. Three other major factors that are changing the face of this region are (1) population changes (both those caused by energy development and those affected by forces other than energy development); (2) changes in the agricultural sector of the region; and (3) changes in the locus of power. We will briefly review these changes in order to set the context in which energy development is occurring.

Perhaps the most obvious and pervasive change that we can observe in the region is that of population growth. Even prior to the past decade, the warmer areas of the west and southwest have experienced some population growth as a function of the expanding recreation and retirement industries. However, the really dramatic changes have occurred post-1970. Prior to the 1970's, the vast majority of rural counties in the Rocky Mountain and Northern Plains states had experienced decades of out-migration of the younger members of their populations as agriculture became less profitable and as opportunities for education and economic advancement became more and more concentrated in the urban communities of the region. The result was that the population of rural counties got both smaller and older.

Over half of the counties of this region that are currently being affected by energy projects actually had smaller populations in 1970 than they did in 1920. And, in some instances, the populations were significantly smaller. However, if we compare their 1980 populations with their 1970 populations, almost all of these counties had reversed the historical pattern of population decline and most have now surpassed the population they had in 1920. Several of the counties had grown by over 100 percent in the decade. Some had experienced annual growth rates of almost 20 percent during their most rapid growth period.

Not only have the populations of these counties gotten larger, they have changed in some other very important ways. For example, they have experienced significant increases in their non-agricultural employment with most of this increase occurring in the mining and construction sectors. As a consequence, their populations have gotten younger. Construction workers, in particular, tend to be young, many are single, and those who are married tend to have young families. They are obviously far more mobile. Many of the construction workers will remain in a community for a brief period and then will move on. This contrasts with the stability of the existing population. For example, more than half the residents of Millard County in Utah have lived there more than 40 years. In addition, the newcomers are much more heterogeneous ethnically, religiously, and culturally. The high level of social and cultural homogeneity of the area in the past is reflected in the fact that most rural counties in Utah have had populations that are over 90 percent Latter-day Saint.

Two major developments have occurred in the agricultural sector of the rural west (and, of American society more generally) that are of particular interest. First, there has been a major change in the size and in the number of farm operations. On a national scale, in 1940 the average farm size was 167 acres, an increase of 14 percent since 1900. In 1960, it had increased by another 67.9 percent. Ten years later it had increased another 30.4 percent. This increase in farm size was accompanied by a significant drop in employment in agriculture (Summers and Lambert, 1981). In the

Rocky Mountain states, the percentage of farm proprietors as a proportion of total employment in the region declined from eight percent in 1970 to six percent in 1977. There was also a decline between 1970 and 1977 in the number of farm workers as a percentage of total employment from about three to two percent. Nationally, there is a trend toward increasing part-time farming and the combination of farm and off-farm work by proprietors. The percentage of farm operators reporting some days off-farm work increased by about 10 percent between 1974 and 1978. In Colorado, Utah, and New Mexico, more than half of all farm operators worked off the farm by 1978 (Summers and Lambert, 1981).

A second development in agricultural organization that is widely documented is the introduction of advanced technology and specialized knowledge. The introduction of sophisticated equipment, chemicals and breeding-hybridization have made it possible to increase productivity while reducing labor intensiveness. The addition of specialized knowledge from agricultural economics, agribusiness management, and horticulture are best incorporated into activities of large corporate entitites, not traditional family farms. Consequently, agriculture is not only supplied and marketed through corporations, but it is also increasingly a corporate activity. By the middle 1970's, these changes had significantly altered the nature of the agricultural enterprise in the west and had created an environment where agriculture could be replaced by energy development as the dominant economic activity in many of the region's rural counties.

The interdependencies that are noted above in the agricultural community have spread to other areas as well and have resulted in "an organization structure of rural society that is increasingly controlled" by forces that are located outside the local community. That is, the increasing centralization of control both within organizations and in networks of organizations means that control leaves rural society. This important change has, in some ways, facilitated the growing dominance of energy companies in these rural counties by changing the key actors who are making the decisions. More importantly, however, energy development has accelerated the shift of control over resources and decision-making from the local community to extra-local actors and agencies.

The most important change in the West, however, and one which is the driving force behind most of the others--particularly that of population change--is the dramatic increase in energy-related activity. Many of the rural communities most significantly affected by energy growth have developed a

stability that has grown out of their particular relationship
to what is often a rather harsh natural environment and is
closely tied to their traditional focus on agricultural and
ranching lifestyles. However, like many of the surrounding
plant and animal communities, they exhibit a fragility that
allows them to be easily and dramatically disrupted by the
change associated with rapid population growth, the importation of large numbers of workers with different cultural
traditions and lifestyles, and the need to respond quickly to
accelerating demands on the community infrastructure and its
public service delivery system (Albrecht, forthcoming).

It is important for us to recognize, however, that
despite the importance of energy development in changing the
face of the west, the actual and potential impacts associated
with these large-scale development of our energy resources
was of relatively little concern until the 1970's (Metz,
1980). In recent years, we have observed significant changes
for a number of reasons. Most important among these are the
following:

1) An increasingly environmentally aware and concerned
citizenry. It was only a little more than a decade ago that
a broad array of talk shows, teach-ins, symposia, celebra-
tions, and demonstrations began to introduce millions of
Americans to their own culpability, as consumers and voters,
for the state of the environment (Courrier, 1980). While the
initial focus of this increasing environmental concern was
directed toward what Dunlap and Catton have called the twin
jaws of a very important vise (Dunlap and Catton, 1979)--
resource depletion and pollution--additional concern was soon
directed toward some of the manifest as well as latent consequences of policy decisions that would lead to the exploitation of huge supplies of abundant energy resources in the
western portion of the United States.

2) The increasingly remote siting of power plants and
other large energy facilities due to increasingly restrictive
environmental laws and regulatory siting criteria. Part of
the opposition to the construction of the huge Kaiparowitz
Power Plant in Southern Utah and to other similar projects
elsewhere resulted from the feeling that the less populated
states were the dumping grounds for California waste. Since
California power companies could not meet restrictive environmental regulations in their own state without the expenditure of much larger sums for pollution abatement, the move
was toward building the plants in other states and then
transporting the relatively clean product--electricity--to
their own.

3) Finally, there has been an increasing recognition on the part of industry that workforce turnover problems that were becoming so common in unattractive energy boomtowns were becoming incredibly costly. This relationship has accelerated industry involvement in socio-economic impact assessment and monitoring. More will be said about this at a later point.

Socio-Cultural Impacts
of Energy Development

The previous chapter has reviewed what appear to be the major socio-cultural impacts that have been and will be associated with large-scale energy development in the West. It should be obvious to all of us that the impacts of these changes cannot be simply labeled either "good" or "bad." In part, our reaction will be based on the perspective we bring to the problem; local business leaders who stand to gain from an energy-development project will react very differently than will retired persons on fixed incomes who will be negatively affected by dramatic increases in costs of housing and goods and services. This forms part of the basis for the argument that western energy development has created an important paradox for area residents. The benefits it brings in the form of new employment and income opportunities have been long sought after; however, the changes in local lifestyles that accompany project-induced growth are viewed with skepticism and anger by many long-time residents.

While it would indeed be a mistake to label the changes either "good" or "bad" without further qualifying these terms, there are some interesting features of what is happening that require special attention. These features distinguish comtemporary changes from anything we have experienced since the settlement of this part of the country. They include (Cortese and Jones, 1977):

1. The rapidity and scale of population growth surpasses what has occurred before. Communities are now doubling and tripling their populations in only a few years.

2. The pervasiveness of the phenomenon is unique in that hundreds of towns in several states are being affected at the same time.

3. Perhaps most importantly, the new energy boomtowns are not being created in a wilderness but are, in fact, long-established, stable agricultural communities whose character will shift dramatically as a result of the development.

4. Due to modern automobile transportation which allows commuting from outlying areas, each large-scale construction job tends to affect not one but several communities.

5. Since many of the projects involve the construction of large power plants requiring large but temporary labor forces, a bust is built into the boom.

As the previous chapter notes, during the past decade there has developed a large and growing literature that has sought to clarify the important social and human consequences of the changes that we have identified. Much of the research has focused on the problems of providing services to the rapidly increasing population (Freudenburg, 1978). Other studies have focused on some of the most important social problems associated with rapid growth. Several early studies argued that these communities experience major increases in social problems--crime rates increase, drug and alcohol abuse problems become more prevalent, child and spouse abuse becomes more serious, marital and mental health problems increase, and so on (see Kohrs, 1974, for example). More recent papers have argued that many of these problems are not as real as was initially assumed (see Wilkinson, Thompson, Reynolds, and Ostresh, 1982). Nevertheless, there are important changes that have occurred and it is to one of these that we now turn.

Some Important Community Impacts
of Rapid Industrialization

The sections that follow will compare what is happening in rapidly growing rural communities of the West with the processes of industrialization and modernization that have occurred in other areas of the country in earlier periods of our history. The argument is that the disruptions and stress that we do observe are a function of the process of change and that if we understand this process, we will have a much better grasp of what is going on around us (see Albrecht, 1980).

The effects of the forces of urbanization and industrialization on the social fabric of communities as well as on the quality of life of individual community residents has long been of concern to social scientists (see, for example, Wirth, 1938; Axelrod, 1956; Kasarda and Janowitz, 1974). Kasarda and Janowitz (1974) note that two models have come to dominate the work of social scientists in this area. They refer to the first as the linear model because of the assumption that linear increases in population size and density are the primary factors that influence patterns of social behavior. As they note, this model clearly has its intellectual roots in Toennies' (1887) concepts of gemeinschaft and gesellschaft. "In this view, urbanization and industrialization alter the essential character of society from that based on communal attachments to an associational basis" (p. 328). Urbanization and industrialization are thus assumed to lead to increased population size and density which, in turn, lead to a breakdown in primary group ties, decreasing attachment to community, greater reliance on secondary institutional supports as opposed to family and kin, and so forth (Wirth, 1938).

The second model grew out of the work of other early University of Chicago sociologists such as Robert E. Park and Ernest W. Burgess (1921, 1925) and finds strong empirical support in the research of Axelrod (1956), Bell and Boat (1957), and Adams (1968), among others. This model, called the systemic model by Kasarda and Janowitz, recognizes extensive primary group ties and informal networks in the urbanized community that are quite contrary to the notion of gesellschaft. "The local community is viewed as a complex system of friendship and kinship networks and formal and informal association ties rooted in family life and on-going socialization processes" (p. 329).

The preponderance of empirical evidence supports the second model. For example, Kasarda and Janowitz (1974) found that residents of communities characterized by increased size and density did not exhibit weaker bonds of kinship and friendship. Neither did location in such communities result in the substitution of secondary for primary and informal ties nor significantly weaken local community sentiments. However, one finds in the work of Kasarda and Janowitz an explicit focus on a variable that was often ignored in earlier work in this area--length of residence. These researchers found that the best predictor of community attachments and the development of social bonds was the amount of time the respondent had resided in the community.

The importance of this factor is evidenced by another line of research which clearly indicates that communities characterized by rapid population growth and change do frequently experience increased levels of social pathology, a breakdown of primary group ties and controls, and other related social problems. This was illustrated in Shaw and McKay's (1942) classic study of juvenile delinquency in Chicago. In discussing the rapid growth that occurred in Chicago around the turn of the century, they note,

> This. . .rapid territorial expansion and geometric
> increase in the population of Chicago implies

marked changes in the areas within the city and a
rate of mobility quite unknown in stable communities. Likewise, the influx of great numbers of
people of such widely different social and cultural backgrounds implies not only lack of
homogeneity but also disorganization and reorganization affecting a large proportion of the
population (p. 25).

In describing the social characteristics of those areas
of Chicago that exhibited the highest rates of juvenile
delinquency, Shaw and McKay note:

Certain areas. . .lack the homogeneity and conti-
nuity of cultural traditions and institutions
which are essential to social solidarity, neighborhood organization and an effective public
opinion. . . .The economic insecurity of the
families, the tendency for the family to escape
from the area as soon as they prosper sufficiently
to do so, all combine to render difficult, if
not impossible, the development of a stable and
effective form of neighborhood organization in
these sections of the city (1942: 110).

Perhaps the problem is that we are dealing with two
rather different issues. In terms of the classic characteristics of gesellschaft, urbanized and industrialized areas
may not be particularly different from more rural communities
once they have become stabilized. That is, individual residents do feel a certain degree of attachment to the community, they do maintain significant informal and primary ties
with kin and friends, and so on. At the same time, communities or areas of communities characterized by the rapid social change that accompanies urbanization and industrialization do experience at least a temporary breakdown in many of
the traditional social support mechanisms that contribute to
such things as community stability, individual identification
with community, and quality of life. Thus, areas of cities
like Chicago that were experiencing rapid social change including growth, changing ethnic character resulting from the
influx of European immigrants, southern blacks, and so on,
were also characterized by high rates of juvenile delinquency, crime, and other important social pathologies. On
the other hand, areas characterized by less social change had
the time to stabilize, individuals re-established important
primary ties with kin and neighbors, came to identify with
community, and developed an overall more stable character
that, in turn, was reflected in much lower rates of crime and
related social problems. Some empirical support for this is

found in the work of Shaw and McKay (1942). Areas of Chicago where blacks have resided for the longest period of time have exhibited some stabilization and have experienced decreasing rates of juvenile delinquency. On the other hand, areas experiencing more recent black invasion have continued to exhibit higher delinquency rates.

When one compares the Shaw and McKay analysis of the disruptive consequences of rapid change with a recent discussion of a western community being impacted by energy development, there are some rather startling parallels. As noted by Freudenburg, et al. (1977: 4-5):

Over the four or five generations that these towns have been inhabited, the residents have developed a fairly impressive set of informal mechanisms-- or "natural systems," if you will--for performing social functions and generally taking care of one another. These mechanisms tend to be of the sort that sociologists can find nearly everywhere (to name a few noteworthy examples, they are ways of controlling deviance, socializing the young, and taking care of the communities' weaker members and/or those in need or under stress).

Yet, in the boomtowns,

In what is probably the most characteristic single consequence of the large-scale impact process, these rather finely-tuned (and surprisingly delicately-balanced) arrangements are simply blown apart--scattered to the four winds by the sudden arrival of more new people than can be contained within them. The process requires no plotting, no nastiness--only numbers. The result is that a people who once took care of one another in a naturally-evolving and in fact almost automatic way--for they are often not even aware of doing so--are suddenly left with some very important machinery that's simply inoperative.

These communities clearly are experiencing the processes of industrialization and urbanization in ways that have serious consequences for their ability to maintain their social fabric and a satisfactory quality of life for their residents. The informal mechanisms that supported and sustained area residents in the past are being broken down and while they will eventually be replaced by other mechanisms in the future, the period before that will be one of serious community crisis.

In other words, the changes in these communities are very real and parallel changes that occurred in the industrializing cities of yesteryear. The magnitude of the social problems that are being experienced is indicative of the fact that they are not simply a result of population growth but that significant social changes are occurring.

A Brief Research Example

The importance of this community level change can be illustrated from a recent project that compared attitudes, perceptions, and community interaction patterns of residents in three communities that have rather different histories of energy development: Price, Utah, and Craig and Gunnison, Colorado. At the time of the research, Craig had recently gone through a rapid growth period from the construction of a large coal-burning power facility in the area. Price was beginning another period of rapid growth from the expansion of coal mining operations in the area and from the construction of a large power plant in neighboring Emery County. Gunnison had experienced little, if any, recent growth but was anticipating major growth from coal development in that area. Data were collected from a random sample of residents in the three communities and from proprietors of all local business establishments through the use of mailed questionnaires. (For a more detailed discussion of the study and the methods that we employed, see England and Albrecht, 1981).

We began our analysis of the communities with the assumption that the tangible forces involved in what the sociologist calls "community" have important implications for the everyday quality of life of individuals. We also proposed that many of the social problems associated with large-scale energy development in rural communities of the West are directly or indirectly attributable to the breakdown of "community"; that is, as a direct function of rapid growth and the diversification of the local economy and the composition of the local population, the community loses, at least temporarily, its ability to provide its inhabitants such things as a sense of place, purpose, identity, personal worth, and so on.

Our attempts to assess this question have been, of necessity, indirect and have relied on the utilization of indirect measures. We do know that these two impacted communities have experienced increasing occurrences of social problems and personal pathologies reflected in higher crime and delinquency rates and increased frequency of drug and alcohol abuse during the early stages of industrialization. Our current data set, however, does not allow us to measure the direct effect of such things as sense of community on these rates. However, if the lower rates of pathology that occurred in the past can be attributed to the operation of "natural systems" that emerge in communities, then length of residence becomes a critical factor and we would anticipate that new residents would have less commitment to, less involvement in, and less of a sense of community than, older residents. This, in turn, should result in their expression of less happiness and less satisfaction with life than older residents. It should also result in fewer social constraints on non-normative behavior and fewer informal support mechanisms which traditionally aid people in responding to the problems and difficulties of life.

It is obvious from our data that length of residence does have a clear and direct effect upon community attitudes and sentiments. This can be seen from a brief examination of responses to the questions we asked. For example, we find a strong positive relationship between length of residence and community sentiments. When asked to rate their community from "best possible community" to "worst," 50 percent of those living in the area more than five years gave a "best" rating (score of 8, 9, or 10). On the other hand, of those living in the community less than a year, 29 percent gave it a "best" rating.

A clear majority of all residence groups defined their community as "home". However, this varied from 95 percent of the long-time residents to 78 percent of the newer residents. Similarly, nearly twice as many of the over-five-year residents indicated that they were "very interested" in things that go on in their community as compared with those who had lived there under a year. The old-time residents were also much more likely to feel that they had a voice in what goes on in their community and that their views were adequately represented. Forty-nine percent of those living in the community over five years felt adequately represented, while this was true of just 22 percent of the newer residents.

It has been argued that the relatively temporary state of many of the construction workers who move into the boom towns of the West effectively inhibits their interest and involvement in the local community. That is, since this will not be their permanent home, why should they expend a great deal of energy to become involved in community affairs and to make the community a better place to live? Bicker (1974) found that commitment to the welfare of the community goes hand-in-hand with long-time residence and a position of higher social status. Similarly, Doran, Duff, and Gilmore (1974) found that many newcomers live on the fringe of the

community, often in temporary trailer settlements, and so
never become highly integrated into the community or participate to a high degree in its affairs. Our findings generally
support these conclusions: newcomers feel less positively
about the community as a place to live, are less likely to
consider it home, are less interested in community affairs,
and are less likely to feel that they have a voice in the
community or that it represents their views.

The systemic model discussed by Kasarda and Janowitz
(1974) also implies that local friendship, kinship, and associational ties contribute to more positive community sentiments. We also proposed that the development of these types
of bonds would be directly related to length of residence in
the community.

The data strongly support the link between friendship
and kinship bonds and length of residence. For example, when
length of residence was compared with percent of friends who
reside in the local community, 77 percent of those living in
the community reported that half or more of their friends
resided in the same town. On the other hand, only 27 percent
of those residing in the community less than a year said half
or more of their friends resided in that same town. The two
intermediate length of residence categories fell about half-
way between these two extremes.

A much smaller percentage of each of the groups reported
that half or more of their relatives resided in their community. Nevertheless, the effect of length of residence was
strongly positive. For example, while just six percent of
those in the community for less than a year said that half or
more of their relatives lived there, this was true of over
five times as many (33 percent) of the long-time residents.

To this point in our analysis we have found that newcomers to the rapidly-industrializing communities that were
studied exhibit less favorable attitudes toward the community, are less interested in its affairs, and feel less certain that it represents their feelings and concerns than old-
timers. Additionally, they are less likely to report that
significant percentages of their close friends and relatives
reside nearby.

When we turn to an examination of the respondent's
participation and involvement in formal organizations and
associations in the community, we observe some most interesting differences. For work organizations, organizations
associated with community affairs, and political organizations, there is a direct linear relationship between level of

participation and involvement for the first three length of residence categories. That is, those who have been in the community less than one year participate less than those who have been there one-to-two years and those who have been there one-to-two years participate less than those who have been in the community three-to-five years. However, those living in the community more than five years exhibit a lower overall level of participation than any of the other three groups, including the recent newcomers. The obvious factor here is age. Our old-timers tend to be significantly older than the other three groups and so their decline in the level of participation in formal associations is not unpredictable.

On the other hand, the newcomers exhibit the highest level of participation in educational, church, and civic organizations. This seems to reflect two basic factors: (1) in terms of educational associations, the newcomers tend to reflect a fairly high level of concern about the quality of local schools and are committed to improving the schools because of the consequences of this for their own children; (2) the newcomers exhibit higher levels of education, income, and occupational status. Therefore, one would expect a high level of participation in certain types of organizations and associations. Particularly the management level people who come in to run the power plants and construction operations should begin to show up fairly early as members of boards of civic and related organizations.

However, there is one other very important factor involved here. Each respondent was asked to indicate if his or her participation in the above organizations occurred within or outside the local community. A significant percentage of the participation of newcomers--ranging from 75 percent for work organizations to 29 percent for church organizations--occurred outside the local community. Except for work organizations, no more than eight percent of any of the three other groups reported their participation to be outside the local community.

The overall pattern here, then, confirms that reported for community sentiments and friendship and kinship bonds. Even when newcomers report higher levels of participation in formal organizations and associations, a significant percentage of this occurs at the extra-local level. In other words, their efforts and interests are focused to a large extent outside the local community and their overall involvement within still tends to be quite limited.

In sum, our data show that newcomers to the communities studied do exhibit weaker friendship and kinship bonds and lower levels of participation in formal and informal local organizations and associations. We feel that this lack of "community imbeddedness" does contribute importantly to the social and personal problems that are being experienced in these communities. Many of the immigrant families are young and have young children. Because of inadequate supplies of local housing, they are often forced to live in unattractive fringe trailer court settlements without lawns and other amenities. When problems occur (i.e., the birth of a new child, work problems, husband-wife disagreements, etc.), local social support mechanisms tend to be inadequate-- family members tend not to live locally, fewer friendship bonds have been established, and so on. The consequence, in many instances, is likely to be increased drinking, more serious marital discord, depression and suicide. It is not likely that these rates would be seriously affected by the construction of new schools or by the completion of the local sewer system. They are, however, likely to be changed by the emergence of a sense of community which included greater friendship and kinship bondedness and increased participation in and commitment to local institutions. Perhaps, more than anything, else, longevity will contribute to the development of the latter.

Conclusion

I will conclude with some brief observations about the role that the social science community, in particular, must play in the events that are occurring in the west. Human values, social institutions like the family, the church, government, education, and the economy, cultural symbols, norms, roles, social class, and power, have been among the most important social science concepts that we have used to understand human social behavior. It requires little imagination to begin to realize the importance of some of these same social science concepts in understanding the important changes that have occurred and are occurring in relationship to our natural environment and the changing rural west. However, the really effective application of our knowledge may require that we change some of the ways we have traditionally thought about our world. As Humphrey and Buttel (1982: 1) have recently noted: ". . .assumptions that energy-intensive industrial development is the natural end point of a universal process of social evolution and modernization must be cast aside if [we] are to break out of [our] collective celebration of Western social institutions." These authors also note that

Not only are we functionally linked to biophysical processes, as are other forms of life, but we are also socialized into normative ways of acting in relation to

that environment. Americans, for example, once
believed that factory smoke meant progress, and
industry willingly polluted the air in cities
such as Chicago and Pittsburgh in the pursuit of
economic growth. Later we found that air pollution increased human mortality and levels of
chronic disease. This medical knowledge became
an important impetus for changing the normative
structure of society with regard to air quality
in the early phase of the environmental movement
in the United States (Humphrey and Buttel, 1982: 3).

There is much that we do not know about the social and
human consequences of some of the important changes that are
occurring. We do have some obligation, however, to continue
to ask hard questions and to make sure that an honest consideration of impacts on the human community becomes a part of
the decision-making equation. We must continue to direct our
attention toward the important social and cultural changes
that are occurring in this area if only so that through the
information obtained, communities which still have some ability to decide whether or not they want to encourage energy-
related growth may be made aware of the consequences--both
positive and negative--of that choice (Cortese and Jones,
1977: 87). The research agenda this provides is both ex-
citing and challenging.

References

Adams, Bert N.
    1968   Kinship in an Urban Setting. Chicago: Markham.

Albrecht, Stan L.
    1980   "Social Participation, Community Attachment, and
            Quality of Life in the Rapidly Industrializing
            Rural Community." Paper presented at the Fifth
            World Congress of Rural Sociology, Mexico City,
            Mexico.

    forth- "Socio-Cultural Factors," in Mohan K. Wali (ed.),
    coming Mining Ecology. London: Academic Press.

Axelrod, Morris
    1956   "Urban Structure and Social Participation." Amer-
            ican Sociological Review 21 (February): 13-18.

Bell, Wendell, and Marion D. Boat
    1957   "Urban Neighborhoods and Informal Social Rela-
            tions." American Journal of Sociology 62 (Janu-
            ary): 391-98.

Bickert, C. von E.
    1974   The Residents of Sweetwater County, Wyoming: A
            Needs Assessment Survey. Denver, Colorado: Denver
            Research Institute.

Cortese, Charles F., and Bernie Jones
    1977   "The Sociological Analysis of Boom Towns." Western
            Sociological Review 8: 76-90.

Courrier, Kathleen (ed.)
    1980   Life After '80: Environmental Choices We Can Live
            With. Andover, Mass.: Brick House Publishing
            Company.

Doran, S., M. K. Duff, and J. S. Gilmore
    1974   Socio-Economic Impacts of Proposed Burlington
            Northern and Chicago Northwestern Rail Line in
            Campbell-Converse Counties, Wyoming. Denver,
            Colorado: Denver Research Institute.

Dunlap, Riley E., and William R. Catton, Jr.
    1970   "Environmental Sociology: A Framework for Analy-
            sis," in T. O. Riordan and R. C. d'Arge (eds.),
            Progress in Resource Management and Environmental
            Planning, Vol. 1. Chicester, England: John Wiley.

England, J. Lynn, and Stan L. Albrecht
    1981   "Social Participation, Community Attachment, and
            Quality of Community Services in the Rapidly Indus-
            trializing Community." Paper presented at the
            Annual Meetings of the American Sociological Asso-
            ciation, Toronto, Canada, August.

Freudenburg, William R.
    1978   "Toward Ending the Inattention: A Report on Social
            Impacts and Policy Implications of Energy Boomtown
            Developments." Paper presented at the meetings of
            the American Association for the Advancement of
            Science, Washington, D. C.

Freudenburg, William R., et al.
    1977   "Subjective Responses to an Energy Boomtown Situ-
            ation: A Preliminary Report on Research in Prog-
            ress in Four Western Colorado Towns." Paper
            presented at the meetings of the American Sociolog-
            ical Association, Chicago, Illinois.

Humphrey, Craig R., and Fred R. Buttel
    1982   Environment, Energy, and Society. Belmont, Ca.:
            Wadsworth.

Kasarda, John D., and Morris Janowitz
1974 "Community Attachment in Mass Society." American Sociological Review 39 (June): 328-39.

Kohrs, ElDean V.
1974 "Social Consequences of Boom Growth in Wyoming." Paper presented at the meeting of the Rocky Mountain American Association for the Advancement of Science, Laramie, Wyoming, April.

Metz, William C.
1980 "The Mitigation of Socio-Economic Impacts by Electric Utilities." Public Utilities Fortnightly, September, 11, 1980: 3-11.

Park, Robert E., and Ernest W. Burgess
1921 Introduction to the Science of Sociology. Chicago: University of Chicago Press.

1925 The City. Chicago: University of Chicago Press.

Shaw, Clifford R., and Henry D. McKay
1942 Juvenile Delinquency and Urban Areas. Chicago: University of Chicago Press.

Summers, Gene F., and Virginia Lambert
1981 "Regional Analysis of Community Changes: Energy Impacts in the 1970's." Research report prepared for Mountain West Research, Inc., Billings, Montana.

Toennies, Ferdinand
1887 Gemeinschaft and Gesellschaft. Leipzig: Feus's Verlag.

Wilkinson, Kenneth P., James G. Thompson, Robert R. Reynolds, and Lawrence M. Ostresh
1982 "Local Social Disruption and Western Energy Development: A Critial Review." Pacific Sociological Review (forthcoming).

Wirth, Louis
1938 "Urbanism as a Way of Life." American Journal of Sociology 44 (July): 3-24.

Steve H. Murdock, F. Larry Leistritz,
Rita R. Hamm, Sean-Shong Hwang

14. An Assessment of the Accuracy and Utility of Socioeconomic Impact Assessments

Socioeconomic impact assessments (Finsterbusch and Wolf, 1981; Wolf, 1980; Murdock and Leistritz, 1979) have played an increasingly important role in the environmental assessment process (Council on Environmental Quality, 1973; 1976; 1978; 1980) requiring substantial investments by public and private concerns. Such assessments have come to form major information bases for public facility planning and for project management, mitigation and monitoring. Socioeconomic assessments have thus come to play not only a legislatively required role but a pragmatic service role as well.

Despite the increasing number of users of impact assessment information and the growing costs for the completion of impact statements, many questions concerning the utility and validity of impact information have not been addressed for either impact statements in general or socioeconomic impact statements in particular. Although some attention has been given to examining the uses of impact information and to describing how it can be presented more effectively (Council on Environmental Quality, 1976; 1978; Denver Research Institute, 1979), few attempts have been made to systematically analyze the likely utility of such data for analytical and planning purposes or to assess the level of accuracy of impact assessments and projections.

The absence of such an analysis has resulted, in large part, of course, from the fact that the assessment process was initiated after 1970, and few data sets containing total coverage of socioeconomic characteristics for small areas

The authors wish to acknowledge the assistance of Karen C. Maki in locating and in providing an initial screening of many of the EISs evaluated in this chapter.

are available except from sources such as the decennial census. Thus, the 1980 census data presently being released provide one of the first opportunities to analyze the accuracy of impact assessments systematically by comparing impact projections with actual census data values. Unless such analyses and assessments are completed, the utility of the impact assessment process and its socioeconomic component cannot be established.

This chapter reports the results of an analysis of the socioeconomic components of a sample of environmental impact statements. The accuracy of a sample of projections made as a part of the EIS process is assessed, and the likely utility of such statements for local impact area decision makers is evaluated. The utility and methodological adequacy of socioeconomic impact assessments were assessed by subjecting formally filed impact statements to a critical analytical assessment (to provide an assessment of assessments). The major inadequacies in each statement are thus established, the accuracy of their population projections evaluated and recommendations for improving socioeconomic EISs formulated. This chapter thus provides both an assessment of socioeconomic impact assessments and, on the basis of this assessment, a set of recommendations for improving the utility of, and policy makers' ability to evaluate, socioeconomic impact assessments.

Methodology

The analysis reported here consists of a systematic review of a sample of socioeconomic impact statements. The initial intent was to select a relatively large number of EISs (between 400 and 500), to extract key economic and demographic projections from these for 1980 and then to evaluate the accuracy of these projections against the economic and demographic data provided by the 1980 Census of Population and Housing. Projections of employment, income, business sales, population, and public services for 1980 by project phase and by geographical area or jurisdiction would be evaluated and compared to determine the accuracy of different impact projection methodologies, of assessments for different areas of the nation and for different types of projects (i.e., coal mines versus uranium mines, power plants, etc.). An initial screening of a large number of EISs revealed, however, that the intended analysis could not be completed. That is, despite the mandate in NEPA for social science analyses, many of the statements reviewed had no socioeconomic section at all or the section did not contain sufficient information (or information for identifiable jurisdictions) to allow anything but broad generalizations to be made about their likely accuracy or the nature of the impacts which potentially impacted areas could expect. A more limited sample and analysis (as described below) were thus necessitated.

The EISs included in this analysis consist of 225 EISs for which complete EIS reports and accompaning appendices could be located. Because of the lack of a comprehensive sampling frame from which to select a sample, the representativeness of the 225 EISs cannot be ascertained. A majority were for projects in the Western U.S. with which the authors were familiar and the potential biasing effect of this areal focus must be recognized. However, the general comments of similar reviews for other parts of the U.S. (Krannich, 1981; Finsterbusch and Wolf, 1981) suggest that the findings reported here may be quite generalizable to SIAs completed throughout the United States.

Of the 225 statements, 94 were for coal-related projects (power plants, gasification facilities and mines), 98 were for nuclear or uranium-related projects and the remainder span a wide range of facilities including coal gasification facilities, copper mines, and nuclear waste sites (salt domes). Table 1 presents a summary of statements consulted by project type.

The findings reported are the result of a two-part analysis of these statements. The first part consists of a content examination of the 225 statements. This analysis was initially intended merely to screen EISs for inclusion in the quantitative evaluation. However, the content analysis revealed a number of problems within EISs, providing important insights into the nature of the socioeconomic impact assessment process, so the results are reported here in some detail. The content analysis thus provides a review of the EISs to determine the types of information and problems inherent in SIAs, as well as an assessment of the appropriateness of the demographic projections of the SIAs for the quantitative comparison to 1980 census results. This analysis, especially the latter part, provides insight into the likely utility of such information for planning and decision making.

The second major phase of the analysis consists of an evaluation of the accuracy of the 1980 population projections from the EISs for counties and cities. Although our initial intent was to complete such evaluations for a full range of economic, demographic and public service projections, we were able to obtain adequate data on only the most basic (total) population projections. As noted

Table 1. Environmental Impact Statements Reviewed by Type of Project

Type of Project	Number	Percent
Nuclear Power Plant	79	35.1%
Coal-Fired Power Plant	50	22.2
Coal Mining	38	16.9
Uranium Mining and Milling	15	6.7
Hydroelectric Generating Plant	8	3.6
Coal Gasification Plant	6	2.7
Nuclear Fuel Facility	4	1.8
Oil Shale	4	1.8
Geothermal	3	1.3
Crude Oil Transportation	3	1.3
Salt Domes	3	1.3
Transmission Lines	3	1.3
Coal Slurry Pipeline	2	.9
Alumite Mine	1	.4
Barite Mine	1	.4
Copper Mine	1	.4
Phosphate Mine	1	.4
Salt Mine	1	.4
Solvent Refined Coal	1	.4
Other	1	.4
TOTAL	225	100.0%

below, however, even these limited data provide important
information on the socioeconomic impact assessment process.

Finally, on the basis of the analysis described above,
a set of policy recommendations for improving the
socioeconomic sections of environmental impact statements is
presented. This section of the chapter discusses
recommendations related to the content, timing and areal
coverage of SIAs. The results of each of the three parts of
the analysis are reported below.

Results

Content Analysis

The content analysis of the impact statements revealed
several major characteristics of socioeconomic impact
assessments and of impact statements in general that merit
discussion. Among the most evident of these characteristics
was a lack of concern for the inclusion of data items
absolutely essential to the socioeconomic assessment
process.

Thus, a surprising number of statements contained no
information on the number, type or distribution of the
workforce--the starting point for any socioeconomic
analysis. Thus, ambiguous statements such as the following
were found in many EISs:

Exact numbers of workers required for construction
are not known at this time.

It is not possible to determine with certainty the
number of local workers to be employed at the
proposed project since the number will depend upon
the capabilities, experience, work preferences,
training, and education of the available local work
force as well as the preconstruction unemployment
rate and the amount of other new construction
projects that may arise.

During construction of the Station, about 1000
construction workers will travel to and from the
site. Most of these workers will be drawn from the
region of the site and to a large extent are
already permanent residents. Others will locate in
neighboring communities and in mobile home parks.
Supervisory personnel will be relocated in the
area.

Table 2. Disposition of Environmental Impact Statements by
Type of Disposition

Disposition of EIS	Number	Percent of Total
Contained Insufficient Information	47	20.9%
Accepted for Further Analysis	44	19.6
No Impact or Baseline Projections	39	17.3
No 1980 Projections*	27	12.0
No Impact Projections	12	5.3
No Baseline Projections	11	4.9
Project Site Unspecified and No Baseline Projections	4	1.8
Impact Area Did Not Correspond to Political Jurisdictions and No Impact Projections	26	11.6
Impact Area Did Not Correspond to Political Jurisdictions and No Baseline or Impact Projections	10	4.4
Impact Area Did Not Correspond to Political Jursidictions and No Baseline Projections	3	1.3
Miscellaneous	2	.9
TOTAL	225	100.0%

*Includes twenty-five recent EISs which describe projects
with start-dates after 1980.

In each case, no quantitative estimates of the workforce were provided, and as a result, no quantitative estimates of other socioeconomic impacts could be made.

Even when quantitative estimates were available, the data were often of questionable utility for planning purposes. Table 2 provides a summary of the disposition of each statement for the present analysis. The categories of rejection shown indicate the reasons statements were **not** included in the present analysis—and likewise the difficulty that would be encountered in attempting to use them for impact planning.

As is evident in the data in this table, only 19.6 percent or 44 statements contained sufficient information to allow the accuracy of their quantitative demographic projections to be tested. That is, only 44 contained population projections for 1980 that are identifiable with a jurisdiction for which data on population are produced by the Census Bureau. Although the percentage of acceptable EISs might legitimately be increased somewhat by the exclusion of the 27 EISs which contained projections for dates other than 1980, the results shown in this table do point to major problems evident in the EISs reviewed. Among these, three are particularly acute: (1) a lack of baseline projections, (2) a lack of impact projections and (3) a lack of projections for areal units that are meaningful to decision makers.

Thus as noted in Table 2, eleven statements gave no baseline projections and seven others had multiple difficulties including providing neither data for distinct jurisdictions nor baseline projections. Although a lack of baseline projections is less problematic than the other two deficiencies noted above, it is still an important limitation. Without data on what the population of the area is likely to be without the particular project of interest, it is difficult to interpret the meaning of an additional number of project-related persons. Impact planners must know whether the area's baseline growth is likely to be so high as to already overtax existing services or whether new (project-related) growth will simply support present businesses and services. Unless baseline projections are provided, the context of development cannot be known and effective decision making cannot occur.

The lack of impact projections is significantly more problematic. Many statements did not contain projections of project-related population. That is, no information on the number of new persons that would come to the impact area as a result of the project (or, of course, on how they would be distributed or their characteristics) was provided. Since such information is essential as a basis for projections of public services, fiscal, social, and other types of socioeconomic impacts, the absence of such information means that these statements would be of little utility for decision making as it relates to socioeconomic dimensions.

When the number of statements rejected for analysis as a result of the absence of baseline, impact or a combination of baseline and impact projections are combined, over 27 percent of the statements were rejected for a lack of projections alone. Given the clear mandate for the inclusion of baseline and impact projections in the federal guidelines for EIS preparation (Council on Environmental Quality, 1973; 1978), the absence of such projections (in statements that are "final" EISs) suggests that either the formal review process of such statements is limited or that the socioeconomic component is considered to be of such limited importance that the absence of projections is not an adequate cause for statement revision or rejection.

The third major limitation is that resulting from the failure to provide projections for areal units with planning utility--that is, for jurisdictions that have elected officials empowered to make public decisions (i.e., public service provision, taxation, etc.). Thus a large number of statements (43 or 19.1%) provided either baseline or impact projections, but provided data for areas without a specified project site or for areas that did not correspond to recognizable jurisdictions. The most common tendency, particularly for nuclear facilities, was to provide projections for an area within a certain radius of the project site. These "kill zone" projections cross boundaries of counties, cities and other identifiable jurisdictions and as such cannot be used for planning in specific jurisdictions (though they may be useful in a very general sense). Although providing projections for such "radius" areas may ensure congruence between the impact area used in the socioeconomic assessments and that in other sections of an EIS (e.g., biological, health and safety, etc. sections), it is essential that socioeconomic impact statements be completed for political jurisdictions so that their results can be used for local and state as well as federal decision making and mitigation planning.

Because of these three problems--the absence of impact or baseline projections and the lack of projections for governmental jurisdictions--over 46 percent of all the statements reviewed were not likely to be useful for impact planning. When the number that contained insufficient (in most cases nonexistent) socioeconomic information (20.9%) is added to the number without impact or baseline projections and/or projections for meaningful jurisdictions, 152 statements (or nearly 68%) did not contain adequate socioeconomic information for decision making, at least not local and state level decision making. Although the impact assessment process is intended for federal decision makers, statements could be (as further indicated below) structured to be of maximum utility to local and state decision makers as well as federal decision makers.

As a final part of the content analysis, the variation in the acceptability of impact statements by the type of proposed project was examined. The results of this analysis are shown in Table 3. Of the facilities with a sufficiently large number of cases to merit comparison, the only notable differences are between nuclear and coal-fired power plants. Because of the extensive use of radial projections around nuclear plants, over 97 percent (see Column 5) of the nuclear power plant statements were not likely to be of utility for local and state decision-making purposes, whereas 50 percent of the coal-fired plant statements provided information that was at least partially appropriate for such purposes.

Overall then, the content analysis of these statements suggests that the socioeconomic assessment process, as practiced through 1980, was largely inadequate. From the standpoint of providing useful information for decision making, the total absence of socioeconomic information (in roughly 21% of the EISs) and the failure to provide projections or the failure to do so for decision-making jurisdictions (46%) clearly suggest major inadequacies in such statements. The evaluation of the accuracy of those 44 statements (roughly 20%) that did contain a sufficient level of information is presented below.

The Accuracy of EIS Population Projections

For those 44 statements which provided population projections for 1980 that could be evaluated against 1980 census population counts, a standard analysis of the accuracy of population projections (National Academy of Sciences, 1980; Zitter and Cavanaugh, 1980) was performed. The results of this analysis are reported in Tables 4 through 8.

Tables 4 and 5 show the 1980 census population values and the EIS projected population values for 1980 for

Table 3. Environmental Impact Statements Reviewed: Dispositions for Analysis by Type of Project

Type of Project	Rejected		Accepted		Percent of Project Type Rejected
	Number	Percent	Number	Percent	
Nuclear Power Plant	77	42.5%	2	4.5%	97.5%
Coal-Fired Power Plant	25	13.8	25	56.8	50.0
Coal Mining	38	21.0	0	0.0	100.0
Uranium Mining and Milling	8	4.4	7	15.9	53.3
Hydroelectric Generating Plant	6	3.3	2	4.5	75.0
Coal Gasification Plant	4	2.2	2	4.5	66.7
Nuclear Fuel Facility	4	2.2	0	0.0	100.0
Oil Shale	1	.6	3	6.8	25.0
Geothermal	3	1.7	0	0.0	100.0

Table 3. Continued

Type of Project	Rejected		Accepted		Percent of Project Type Rejected
	Number	Percent	Number	Percent	
Crude Oil Trans-					
portation	3	1.7	0	0.0	100.0
Salt Domes	3	1.7	0	0.0	100.0
Transmission Lines	3	1.7	0	0.0	100.0
Coal Slurry Pipe-					
line	2	1.1	0	0.0	100.0
Alumite Mine	0	0.0	1	2.3	0.0
Barite Mine	1	.6	0	0.0	100.0
Copper Mine	1	.6	0	0.0	100.0
Phosphate Mine	0	0.0	1	2.3	0.0
Salt Mine	1	.6	0	0.0	100.0
Solvent Refined					
Coal	1	.6	0	0.0	100.0
Other	0	0.0	1	2.3	0.0
TOTAL	181	100.0%	44	100.0%	

Table 4. Projected and Actual (Census) 1980 Populations for Counties by State

State	Counties	Census Population 1980	EIS Projected Population 1980	Numerical Difference	Percent Difference
Arizona	Navajo	67,709	55,000	-12,709	- 18.8
Arkansas	Independence	30,147	24,564	- 5,583	- 18.5
Colorado	Jackson	21,646	23,000	1,354	6.3
	Delta	21,225	21,100	- 125	- .6
	Garfield	22,514	33,000	10,486	46.6
	Garfield	22,514	25,300	2,786	12.4
	Gunnison	10,689	9,400	- 1,289	- 12.1
	Mesa	81,530	91,950	10,420	12.8
	Mesa	81,530	67,616	-13,914	- 17.1
	Moffat	13,133	15,182	2,049	15.6
	Montrose	24,352	22,900	- 1,452	- 6.0
	Ouray	1,925	2,200	275	14.3
	Pitkin	10,338	17,050	6,712	64.9
	Rio Blanco	6,255	14,974	8,719	139.4
	Routt	13,404	16,366	2,962	22.1
	Regional Area	110,299	137,750	27,451	24.9

Table 4. Continued

State	Counties	Census Population 1980	EIS Projected Population 1980	Numerical Difference	Percent Difference
Georgia	Bibb	151,085	166,000	14,915	9.9
	Carroll	56,346	54,000	- 2,346	- 4.2
	Coweta	39,268	40,000	732	1.9
	Heard	6,520	5,000	- 1,520	- 23.3
	Jones	16,579	16,500	- 79	- .5
	Monroe	14,610	12,000	- 2,610	- 17.9
	Troup	50,003	44,600	- 5,403	- 10.8
Idaho	Bannock	65,451	75,253	9,802	15.0
	Bear Lake	6,931	5,950	- 981	- 14.1
	Bingham	36,489	35,061	- 1,428	- 3.9
	Caribou	8,695	9,501	806	9.3
	Franklin	8,895	7,166	- 1,729	- 19.4
	Oneida	3,258	2,638	- 620	- 19.0
	Power	6,844	5,748	- 1,096	- 16.0
Minnesota	Aitkin	13,404	9,300	- 4,104	- 30.6
	Carlton	29,936	28,870	- 1,066	- 3.6
	Cook	4,092	3,330	- 762	- 18.6
	Itasca	43,006	34,720	- 8,286	- 19.3
	Koochiching	17,571	17,760	189	1.1

Table 4. Continued

State	Counties	Census Population 1980	EIS Projected Population 1980	Numerical Difference	Percent Difference
Minnesota (Cont.)	Lake	13,043	12,950	– 93	– .7
	St. Louis	222,229	226,880	4,651	2.1
Montana	Big Horn	11,096	10,694	– 402	– 3.6
	Big Horn	11,096	11,602	506	4.6
	Custer	13,109	13,523	414	3.2
	Powder River	2,520	2,320	– 200	– 7.9
	Rosebud	9,899	9,720	– 179	– 1.8
	Rosebud	9,899	9,900	1	0.0
	Rosebud	9,899	9,374	– 525	– 5.3
	Treasure	981	1,220	239	24.4
Nevada	Humboldt	9,434	8,001	– 1,433	– 15.2
	Lander	4,082	3,382	– 700	– 17.1
New Mexico	McKinley	54,950	64,750	9,800	17.8
	Rio Arriba	29,282	28,100	– 1,182	– 4.0
	Sandoval	34,799	24,400	–10,399	– 29.9
	San Juan	80,833	68,700	–12,133	– 15.0
	San Juan	80,833	89,282	8,449	10.4
	San Juan	80,833	66,220	–14,613	– 18.1
	Valencia	60,853	49,900	–10,953	– 18.0

Table 4. Continued

State	Counties	Census Population 1980	EIS Projected Population 1980	Numerical Difference	Percent Difference
New York	Cayuga	79,894	82,000	2,106	2.6
	Columbia	59,487	64,000	4,513	7.6
	Greene	40,861	45,000	4,139	10.1
	Jefferson	88,151	89,000	849	1.0
	Lewis	25,035	24,000	- 1,035	- 4.1
	Madison	65,150	73,000	7,850	12.0
	Oneida	253,466	285,000	31,534	12.4
	Onondaga	463,324	518,000	54,676	11.8
	Ontario	88,909	91,000	2,091	2.4
	Oswego	113,901	122,000	8,099	7.1
	Seneca	33,733	38,000	4,267	12.6
	Ulster	158,158	170,000	11,842	7.5
	Wayne	85,230	93,000	7,770	9.1
North Carolina	Davidson	118,393	113,162	- 5,231	- 4.4
	Davie	22,806	24,599	1,793	7.9
	Rowan	99,186	97,675	- 1,511	- 1.5
Oklahoma	Regional Area	274,716	263,301	-11,415	- 4.2

Table 4. Continued

State	Counties	Census Population 1980	EIS Projected Population 1980	Numerical Difference	Percent Difference
Utah	Beaver	4,378	7,887	3,509	80.2
	Carbon	22,179	22,490	311	1.4
	Carbon	22,179	23,030	851	3.8
	Emery	11,451	11,690	239	2.1
	Emery	11,451	9,770	- 1,681	- 14.7
	Garfield	3,673	3,940	267	7.3
	Garfield	3,673	4,540	867	23.6
	Grand	8,241	7,978	- 263	- 3.2
	Iron	17,349	17,672	323	1.9
	Kane	4,024	3,000	- 1,024	- 25.4
	San Juan	12,253	17,373	5,120	41.8
	Sevier	14,727	14,570	- 157	- 1.1
	Wayne	1,911	2,660	749	39.2
	Regional Area	15,949	14,990	- 959	- 6.0
Washington	Benton	109,444	91,841	-17,603	- 16.1
	Franklin	35,025	32,780	- 2,245	- 6.4
Wyoming	Campbell	24,367	23,330	- 1,037	- 4.3
	Carbon	21,896	21,738	- 158	- 0.7
	Carbon	21,896	23,733	1,837	8.4

Table 4. Continued

State	Counties	Census Population 1980	EIS Projected Population 1980	Numerical Difference	Percent Difference
Wyoming (Cont.)	Carbon	21,896	24,454	2,558	11.7
	Converse	14,069	15,490	1,421	10.1
	Converse	14,069	21,037	6,968	49.5
	Converse	14,069	23,958	9,889	70.3
	Converse	14,069	21,662	7,593	54.0
	Fremont	40,251	34,815	- 5,436	13.5
	Johnson	6,700	7,470	770	11.5
	Lincoln	12,177	8,676	- 3,501	28.8
	Sheridan	25,048	23,839	- 1,209	4.8
	Sheridan	25,048	18,370	- 6,678	26.7
	Sweetwater	41,723	47,305	5,582	13.4
	Sweetwater	41,723	49,150	7,427	17.8
	Sweetwater	41,723	34,408	- 7,315	17.5
	Uinta	13,021	13,281	260	2.0

Table 5. Projected and Actual (Census) 1980 Populations for Cities by County and State

State	County	Place	Census Population 1980	EIS Projected Population 1980	Numerical Difference	Percent Difference
Arizona	Navajo	Holbrook	5,785	6,500	715	12.4
	Navajo	Winslow	7,921	9,000	1,079	13.6
Colorado	Mofatt	Craig	8,133	13,721	5,588	68.7
	Rio Blanco	Meeker	2,356	7,279	4,923	209.0
	Rio Blanco	Rangely	2,113	5,551	3,438	162.7
	Routt	Hayden	1,720	2,939	1,219	70.9
	Routt	Steamboat Springs	5,098	9,013	3,915	76.8
Georgia	Bibb	Macon	116,860	140,436	23,576	20.2
	Carroll	Carrollton	14,078	16,038	1,960	13.9
	Coweta	Newnan	11,449	13,640	2,191	19.1
	Heard	Franklin	711	700	− 11	− 1.6
	Monroe	Forsyth	4,626	4,080	− 546	− 11.8
	Troup	La Grange	24,204	233,370	209,166	864.2
Nevada	Humboldt	Winnemucca	4,140	4,481	341	8.2

Table 5. Continued

State	County	Place	Census Population 1980	EIS Projected Population 1980	Numerical Difference	Percent Difference
North Dakota	Mercer	Benlah	2,878	4,685	1,807	62.3
Utah	Emery	Castle Dale	1,910	2,320	410	21.5
Utah	Emery	Emery	372	430	58	15.6
Utah	Emery	Ferron	1,718	1,760	42	2.4
Utah	Emery	Huntington	2,316	2,190	- 126	- 5.4
Utah	Emery	Orangeville	1,309	1,320	11	0.8
Washington	Douglas	Bridgeport	1,174	1,840	666	56.7
Washington	Douglas	Mansfield	315	510	195	61.9
Washington	Okanogan	Brewster	1,337	1,780	443	33.1
Washington	Okanogan	Coulec Dam	1,412	2,030	618	43.8
Washington	Okanogan	Okanogan	2,302	2,635	333	14.5
Washington	Okanogan	Pateros	555	850	295	53.1
Wyoming	Carbon	Baggs	433	465	32	7.4
Wyoming	Carbon	Elk Mountain	338	246	- 92	- 27.2
Wyoming	Carbon	Encampment	611	533	- 78	- 12.8
Wyoming	Carbon	Medicine Bow	953	861	- 92	- 9.7
Wyoming	Carbon	Rawlins	11,547	13,336	1,789	15.5
Wyoming	Carbon	Saratoga	2,410	2,165	- 245	- 10.2

Table 5. Continued

State	County	Place	Census Population 1980	EIS Projected Population 1980	Numerical Difference	Percent Difference
Wyoming (Cont.)	Carbon	Sinclair	586	562	- 24	- 4.1
	Converse	Douglas	6,030	16,963	10,933	181.3
	Converse	Douglas	6,030	14,075	8,045	133.4
	Converse	Glenrock	2,736	5,588	2,852	104.2
	Converse	Glenrock	2,736	5,388	2,652	96.9
	Platte	Wheatland	5,816	6,910	1,094	18.8
	Sheridan	Sheridan	15,146	17,086	1,940	12.8
	Sweetwater	Green River	12,807	14,185	1,378	10.8
	Sweetwater	Rock Springs	19,458	29,909	10,451	53.7
	Sweetwater	Rock Springs	19,458	18,452	- 1,006	- 5.5
	Sweetwater	Wamsutter	681	463	- 218	- 32.0
	Uinta	Evanston	6,421	6,995	574	8.9
	Uinta	Lyman	2,284	2,729	445	19.5

Table 6. Projection Error for Counties and Cities, 1980

Area	(N)	Mean 1970–80 Percent Change	Mean Percent Error	Standard Deviation Mean Percent Error	Mean Absolute Percent Error	Standard Deviation Mean Absolute Percent Error
Counties	(104)	40.2	3.9	24.3	15.4	19.2
Cities	(45)	85.5	54.7	134.4	59.8	132.2

Table 7. Projection Error for Counties, 1980 by Size and Percent Population Change, 1970-1980

Population Size, 1970	(N)	Mean 1970-80 Percent Change	Mean Percent Error	Standard Deviation Mean Percent Error	Mean Absolute Percent Error	Standard Deviation Mean Absolute Percent Error
<5,000	(13)	23.6	17.2	47.4	33.3	36.9
5,001 - 10,000	(23)	71.7	9.4	28.5	21.3	20.7
10,001 - 15,000	(15)	36.7	0.9	16.5	9.7	13.1
15,001 - 20,000	(13)	59.9	- 3.4	14.2	10.5	9.8
20,001 - 50,000	(17)	22.7	- 4.5	11.4	10.2	6.4
50,001+	(23)	22.3	3.1	11.3	9.8	6.2

Table 7. Continued

Percent Population Change, 1970–80	(N)	Mean Population Size, 1970	Mean Percent Error	Standard Deviation Mean Percent Error	Mean Absolute Percent Error	Standard Deviation Mean Absolute Percent Error
<10.0	(17)	92,509	5.4	8.0	7.3	6.2
10.1 – 20.0	(18)	67,730	2.5	23.3	14.3	18.3
20.1 – 30.0	(16)	26,394	13.1	38.5	22.8	33.3
30.1 – 50.0	(25)	22,319	– 7.2	12.5	11.5	8.5
50.1 – 100.0	(17)	20,492	0.8	24.3	16.6	17.3
100.1+	(11)	9,274	–20.2	27.7	26.1	21.6

Table 8. Projection Error for Cities, 1980 by Size and Percent Population Change, 1970-1980

Population Size, 1970		(N)	Mean 1970-80 Percent Change	Mean Percent Error	Standard Deviation Mean Percent Error	Mean Absolute Percent Error	Standard Deviation Mean Absolute Percent Error
	<500	(9)	121.0	5.8	33.0	24.9	20.7
501 -	1,000	(10)	125.9	110.6	266.6	112.0	265.9
1,001 -	2,500	(10)	63.2	73.6	69.8	75.6	67.4
2,501 -	10,000	(10)	69.9	44.1	64.1	46.5	62.2
10,001+		(6)	27.7	20.9	16.9	20.9	16.9

Table 8. Continued

Percent Population Change, 1970–80	(N)	Mean Population Size, 1970	Mean Percent Error	Standard Deviation Mean Percent Error	Mean Absolute Percent Error	Standard Deviation Mean Absolute Percent Error
<10.0	(7)	22,591	139.0	320.0	139.5	319.8
10.1 - 30.0	(8)	2,107	28.5	26.8	31.5	22.7
30.1 - 50.0	(6)	4,468	67.5	93.1	68.8	91.9
50.1 - 100.0	(7)	4,441	47.4	45.8	51.1	40.9
100.1+	(17)	1,197	30.7	57.5	40.6	50.5

counties (Table 4) and cities (Table 5) for the areas
projected in the 44 EISs. Since projections for multiple
sites were included in several EISs, the number of
projections for counties (104) and for cities (45) is
greater than the number of EISs.

The data in Tables 4 and 5 clearly suggest a wide
discrepancy between the actual 1980 populations for various
areas and those projected in the impact statements. For
counties (Table 4), for example, 57 of the 104 projections
(or 55%) were in error by more than 10 percent, and 14 (or
13%) were in error by over 25 percent. In addition, it is
evident that the statements were nearly equally likely to
underestimate as overestimate the populations of counties.
That is, 52 percent (54) of the county populations were
overestimated, while 48 percent (50) were underestimated.

For cities (Table 5), the differences between the
actual and EIS projected values are even larger. Thus for
35 (77%) of the 45 city populations projected by the EISs,
the errors exceeded 10 percent and for 19 (42%) the level of
error exceeded 25 percent. In fact, for 6 projections
(13%), the errors exceeded 100 percent. Unlike the
projections for counties, however, the projections for
cities show a consistent pattern of overestimation. Thus
only 9 (or 20%) of the city projections underestimated the
actual 1980 city populations. Overall, then, the
projections for counties and cities vary widely in accuracy
and clearly require careful evaluation.

Finally, it is interesting to note the wide variations
in different projections for the same areas (Tables 4 and
5). Table 4 shows two separate projections for Mesa County,
Colorado—one projection underestimated the county's
population by nearly 14,000 and one overestimated its
population by 10,000; for San Juan County, New Mexico—one
projection overestimated the county's population by 8,000
while another underestimated it by 14,000; and for
Sweetwater County, Wyoming—one projection overestimated the
county's population by 7,000 and one underestimated the
population by 7,000. For cities, the limited number of
duplicate projections makes fewer comparisons possible, but
for Rock Springs (a city with a population of only 19,000 in
1980) the two 1980 projections for the city differed by over
10,000 persons. Clearly such levels of error are
unacceptable and such projections of little utility for
decision making.

Tables 6 through 8 further summarize the accuracy of
the projections provided by the EISs. Table 6 indicates the

mean percentage error (in which positive and negative percentage errors are simply summed with some resultant cancelling of errors [+ and -] reducing the overall level of error) and the mean absolute percentage error (in which the signs of the error are ignored) for counties and cities, and Tables 7 and 8 show the variation in county and city errors by the population size of the areas in 1970 and by the areas' percent change in population from 1970 to 1980.

As is evident from examining Table 6, the mean percent error for counties was only 3.9 percent (because overestimated and underestimated values tended to cancel one another), but the mean absolute percentage error was over 15 percent. For cities the mean and mean absolute percentage error were both over 50 percent. Although one can generally expect that the level of error will increase as the rate of population change in an area increases and will be less for areas with large population sizes (Shryock and Siegel, 1980; National Academy of Sciences, 1980), the data in Tables 7 and 8 suggest that such patterns do not occur for either the county or city projections provided in the EISs. This indicates that the lack of accuracy of the projections is thus not merely a result of the rapid growth or small population sizes of the projected areas, factors that are known to increase the error of estimate for population projections (Shryock and Siegel, 1980).

Overall, then, the assessment of the accuracy of projections for even the small number of statements that contained data adequate for such an analysis is not encouraging. Such characteristics as relatively small population size, rapid growth and reversal of past population trends can be expected to lead to larger projection errors (Pittenger, 1976; Zitter and Cavanaugh, 1980; Shryock and Siegel, 1980). However, Census Bureau estimates for counties are generally seen as being only marginally useful if the mean absolute percentage error is greater than 5 to 10 percent (National Academy of Sciences, 1980), and errors of over 25 percent for counties or cities are nearly always seen as excessive and likely to indicate that the projections are of little practical utility (Isserman, 1977). Given that the mean absolute percentage errors for the EIS projections reported here were over 15 percent for counties and over 50 percent for cities, it is evident that such projections are likely to be of little practical utility for planning and decision making and that the technical adequacy of such projections is clearly questionable.

Summary and Policy Implications

The analysis presented above clearly reveals that the socioeconomic sections of environmental impact statements have severe limitations. Many statements simply ignore the socioeconomic dimensions entirely, others fail to provide necessary baseline or impact projections and still others do not provide data for jurisdictions that are useful for local and state level decision making. Finally, for those that do provide projections for identifiable jurisdictions, the accuracy of their projections is such as to make the use of the projections of questionable value. The analysis presented here is limited, of course, and it is evident that further analyses of a much larger sample of impact statements and for several projection periods (in addition to 1980) and of assessments evaluating multiple dimensions (in addition to population) are essential before any definitive recommendations should be made. However, we believe the findings are sufficiently uniform and pervasive so as to suggest the basis for formulating a set of policy recommendations or guidelines for socioeconomic impact assessments.

On the basis of this analysis and the prevailing literature in socioeconomic assessments (Finsterbusch and Wolf, 1981; Denver Research Institute, 1979; Murdock and Leistritz, 1979; Leistritz and Murdock, 1981) we recommend as initial steps toward improving socioeconomic assessments that the following time, areal and content-related guidelines be implemented.

Time Guidelines

These guidelines relate to the time periods covered by statements and the time and project-phase referents of their reported results:

• All socioeconomic assessments should include baseline and impact projections.

• Projections should be required for each year during project construction, the first year of project operation and for a minimum of every fifth year of the operational period thereafter. Whenever possible, projections should also be made for the first few years after the end of the project's operational life.

• Projected time periods should include census years to allow updating and corrections of projections to be made as the project proceeds through systematic comparisons of census and projected results.

Areal Guidelines

These include guidelines in reference to both of the types of areas to be covered by statements and the coordination of results for various levels of geographical detail:

- Socioeconomic projections should include projections for the key governmental jurisdictions in the impact area.

- The minimum governmental jurisdictions for which projections should be made are counties, cities and school districts.

- The projections for multiple jurisdictions should be coordinated to ensure that the projections at each level are congruent with those at other levels (i.e., the sum of projections for cities and rural parts of counties should sum to county totals and the sum of county projections to those for multicounty regions, etc.).

Content Guidelines

In reference to the content of socioeconomic assessments, we recommend that such assessments provide data on the following basic components and dimensions within each component:

- Economic--including changes in

  1. business activity
  2. personal income
  3. employment (by type--construction, operational, etc.)

- Demographic--including changes in

  1. regional, county, and community population
  2. population settlement patterns
  3. population characteristics

- Community services--including changes in the level, distribution, and quality of

  1. water and sewer services
  2. police and fire services
  3. housing

4. recreational services
5. educational services
6. transportation
7. social services
8. medical services

• Fiscal--including the level and distribution by jurisdiction, type of service, and source of

1. public revenues
2. public costs
3. net fiscal balances

• Social--including changes in

1. social structures
2. social institutions
3. community and individual perceptions
4. conditions of specific groups such as the elderly, minority groups, and those on fixed incomes

In addition, in relation to the content of socioeconomic assessments, we believe that in the analyses of the factors noted above (economic, demographic, etc.), socioeconomic assessment analysts should systematically consider the likely effects of existing and possible changes in three major sets of characteristics which are known to affect the magnitude and distribution of project-related impacts. These are:

• Characteristics of the project

1. location
2. work force size during various project phases
3. length of project phases
4. level of project expenditures in the local area

• Characteristics of the area

1. the likely level of local employment at the project
2. the level of present services and fiscal bases
3. the likely overall attractiveness of local settlement areas
4. the social organization and characteristics of the indigenous population

• Characteristics of new workers

1. nonlocal worker characteristics (including those of

their families)
2. nonlocal worker settlement patterns
3. nonlocal worker service preferences, demands and requirements

Finally, we suggest that, given the state of the art:

• Socioeconomic assessments should include a range of projections for each development alternative (including the no-action alternative)

These recommendations have often been made by ourselves (Murdock and Leistritz, 1979; Leistritz and Murdock, 1981) and by numerous other researchers (Denver Research Institute, 1979; Fitzsimmons et al., 1975; Wolf, 1980; Finsterbusch and Wolf, 1981), but the analysis reported here dramatizes the need for their immediate implementation. The methods required by these   recommendations are presently available in readily applicable forms (Denver Research Institute, 1979; Leistritz and Murdock, 1981); it is thus past time that the state-of-the-practice should come to approximate the state-of-the-art. If this occurs, we believe the socioeconomic assessment process may yet come to be of utility for decision makers at all levels of government and may yet fulfill the promise of the mandates that created it.

References

Council on Environmental Quality. 1980. **102 Monitor** (December).

_____ . 1978. "National Environmental Policy Act." **Federal Register** 43 (June 9): 112.

_____ . 1976. "Environmental Impact Statements: An Analysis of Six Years' Experience by Seventy Federal Agencies." (March).

_____ . 1973. "Preparation of Environmental Impact Statements." **Federal Register** 38 (August 1): 147.

Denver Research Institute. 1979. **Socioeconomic Impact of Western Energy Resource Development.** Washington, D.C.: Council on Environmental Quality.

Finsterbusch, Kurt and C. P. Wolf. 1981. **Methodology of Social Impact Assessment.** Stroudsberg, Pennsylvania: Dowden, Hutchinson and Ross, Inc.

Fitzsimmons, S. J., L. I. Stuart and C. P. Wolf. 1975. **Social Assessment Manual: A Guide to the Preparation of the Social Well-Being Account.** Washington, D.C.: U.S. Bureau of Reclamation.

Isserman, A. M. 1977. "The Accuracy of Population Projections for Subcounty Areas." **Journal of the American Institute of Planners** 43: 247-259.

Krannich, Richard. 1981. "Socioeconomic Impacts of Power Plant Developments on Nonmetropolitan Communities." **Rural Sociology** Vol. 4 (1): 128-142.

Leistritz, F. Larry and Steve H. Murdock. 1981. **The Socioeconomic Impact of Resource Development: Methods for Assessment.** Boulder: Westview Press.

Murdock, Steve H. and F. Larry Leistritz. 1979. **Energy Development in the Western United States: Impact on Rural Areas.** New York: Praeger Publishers.

National Academy of Sciences. 1980. **Estimating Populations and Income of Small Areas.** Washington, D.C.: National Academy Press.

National Environmental Policy Act. 1970. Public Law, PL-190.

Pittenger, Donald. 1976. **Projecting State and Local Populations.** Cambridge, Mass.: Ballinger Publishing Company.

Shryock, H. S. and J. S. Siegel. 1980. **The Methods and Materials of Demography.** Washington, D.C.: U.S. Bureau of the Census, U.S. Government Printing Office.

Wolf, C. P. 1980. "Getting Social Impact Assessment into the Policy Arena." **Environmental Impact Assessment Review** 1 (March): 27-36.

Zitter, Meyer and Frederick J. Cavanaugh. 1980. "Postcensal Estimates of Population." A paper delivered at the 1980 Annual Meeting of the American Association for the Advancement of Science, Session on the 1980 Census. January 5.

Joseph G. Jorgensen

15. Energy Developments in the Arid West: Consequences for Native Americans

Introduction

Energy-related developments have exercised profound effects on many rural regions of the western U.S., from the Fort Union coal field in Montana to the Grants Mineral Belt in New Mexico. Discussions of these effects in other chapters of this section as well as in the broader scientific literature have generally focused on "Anglo" towns.[1] In this chapter, however, we are interested in Native Americans, or Indians (the latter being the term that, from habit, the author, and "Native Americans," prefer to use).

Indians have much different ways of life from Anglos, and anthropologists have long sought to explain the similarities and differences between Anglo and Indian cultures. In the past decade, particularly since the Arab oil embargo and the formation of OPEC, transnational energy corporations and the federal government have sought to better understand Indians while also seeking to make deals for the energy resources on many reservations.[2] The importance of Indian energy resources should not be underestimated. They surely have not been underestimated by most of the world's leading energy corporations. Nor should the importance of contemporary Indian cultures and their histories be underestimated. But corporation executives, local, state, and federal officials, local entrepreneurs, and laborers on energy projects often do.

Indian tribes own about a quarter of the nation's known strippable coal reserves (perhaps 110 billion tons), four percent of the nation's oil reserves (1 billion barrels), somewhere between 10 and 15 percent of the proven uranium reserves, and very large quantities of the nation's known active geothermal areas, tar sands reserves, natural gas reserves, and oil shale reserves. Of equal importance to the

energy resources are Indian claims to water and water rights,
rights provided to them by treaties and by the U.S. Supreme
Court's Winter's Doctrine of 1908. Water in prodigious quantities are required to mine, transport and convert the energy
resources in the arid west.[3]

The ownership of those resources, however, is different
from ownership as exercised by a corporation, even though all
tribes that adopted and ratified both constitutions and char-
ters under the provisions of the Indian Reorganization Act
of 1934[4] became incorporated. Fuller analyses of the limits
imposed on tribal corporations than can be provided in the
space here are available.[5] In brief, there are definite
legal limitations on each Indian tribe's authority over its
resources, the members of the tribe, their external and internal business affairs, and on their authority over persons
on tribal territory who are not members of the tribe. Any
action or decision made by a tribe can be vetoed by the
Secretary of Interior; the federal government has trust obligations for Indians, their land, and their water, but does
not exercise such authority over privately-owned property,
non-Indian persons, local governments, or state governments.

Given the resources on many Indian reservations in the
West, commentators for the major print and television media,
the federal government, and the energy corporations have
claimed that energy resources on Indian lands hold the potential for significant financial benefits for reservations and
their inhabitants.[6] Whether any "significant" benefits have
been achieved is a perplexing problem to be discussed below.
For the greatest part, the record has not been a positive
one and the promise of prosperity has not been fulfilled.
In the meantine, Indian persons and communities have been
affected in a number of unanticipated ways by existing and
proposed energy developments.

One of the clearest trends is an increase in conflicts.
Tribal members have sued their elected leaders in federal
court over contracts the tribal authorities have signed with
transnational energy corporations.[7] Rank-and-file members
of tribes whose traditional residences and land areas have
been threatened by energy-related developments, have sued
corporations and agencies for failing to analyze the communities, ways of life, and dependencies of those tribal people
on their traditional areas. Residents in small hamlets on
reservations have accused their tribal governments of failing
to protect them from discrimination and to protect their land
from abuse by employees of energy corporations in their
midsts.[8] Households have been involuntarily relocated from
their traditional residences and resources because of the

opening of mines, mills, electricity generating plants, and
the like. In several instances natives have been caused to
sever ties with kinspeople, friends, burial sites, shrine
areas, and expectations that their progeny could reside in
those same areas.[9]

The disruptive consequences of energy development have
also extended beyond matters of conflict. Reciprocity-based
kinship networks have been affected, and expectations have
been changed. It is possible that energy development is not
required to be disastrous for Indians; yet the contexts from
which tribes operate--little capital and meager information--
and the nature of corporate capitalism in energy extraction[10]
have yielded unpleasant consequences in virtually all cases
that have been studied to date.

A Framework for Explanation

The social science research literature is almost exclusively focused on Anglo, i.e., non-Indian communities. Although this is not the proper place to explore the "mind-
sets" of social scientists who have engaged in the research
and analysis of the social consequences of energy developments,[11] it is important to point out that among social scientists, anthropologists (more often than sociologists,
economists, political scientists, or psychological counsel-
ors) conduct research among Indians. Furthermore, anthropologists usually reside in communities and conduct their
inquires over much longer periods than is true in most
studies of Anglo communities. In so doing, they often come
to understand how communities work and the beliefs that are
shared. In general, anthropologists have found that Indians
do not share the "mind sets" of most social scientists nor
those of ranchers, farmers, or entrepreneurs in the arid
west. Social scientists--like engineers, businessmen,
physicians, and a variety of other professionals--are not
isolated from the affects of their own culture.

One particularly powerful aspect of western culture is
the commodity principle, in which people's productive capa-
cities and nature's resources are among the "things" which
can be bought and sold in the market place. Many apparently
"rational" or "scientific" approaches, such as "cost-benefit"
analysis in economics, are actually reflections of this
specific cultural principle. Economics, with its focus on
price, regression analysis, and growth; business, with its
cost-accounting forms; engineering, with its language of
numbers and its assumptions of growth and saturation curves;
and the belief of chemists and physicists that progress
means technological progress also reflect the commodity

principle. Some anthropologists, such as Laura Nader, have pointed out that these examples represent variations on an encompassing mind-set that allows for "acceptable" costs in the interest of developing economic and social programs world-wide. There need be no maliciousness, nor even under-standing of principles that shape their views.

Such ways of looking at the world are not universal or "inherently" proper, however, For example, during the past century students of Indian societies in the west have learned that Indians do not define the places in which they reside nor the spaces where they obtain their livelihoods solely by ownership rights to corporeal property.[12] The evidence from many contemporary Indian societies suggests that reser-vation Indians continue to evaluate land as spaces where livelihoods are obtained, places where present and future generations will reside, and spaces that are part of nature, yet are endowed with spirits that are more than natural and that are of special consequences and meanings to past, present, and future generations.

It is a reasonable simplification to say that Indians tend to have what Anglos consider to be irregular attitudes about land, about the environment of which land is but a part, and about the cultures—past, present, and future—that will live upon the land.[13] Indians mean it when, as Russell Jim, the Yakima tribal councilman, said in relation to using Yakima ceded lands for nuclear generation stations and waste dumps: "from time immemorial we have known a special relationship with Mother Earth....We have a moral duty to help protect Mother Earth from acts which may be a detriment to generations of all mankind." And they also meant it when the Dalton Pass Chapter of the Navajo Tribe framed the following statement in relation to proposed uran-ium exploration (drilling) on their land: "as long as this uranium exploration continues our lives and the lives of future generations are not safe. We will not allow our health, welfare, land, and culture to be destroyed by cor-porate outsiders." The traditionalist Hopi villages who joined that Black Mesa Defense Fund suit to stop the Peabody excavation, expressed their "irregular" conception of the value of land thus: "Hopi land is held in trust in a spiritual way for the Great Spirit, Massau'u....The area we call "Tukunavi" [which includes Black Mesa] is part of the heart of our Mother Earth....This land was granted by a power greater than man can explain. Title is vested in the whole makeup of Hopi life....The land is sacred and if the land is abused, the soundness of Hopi life will disappear and all other life as well." Attitudes similar to the pre-ceding have been confirmed repeatedly in surveys of modern Indian communities.[14]

Those who plan and direct America's energy future, however, have "mind-sets" of their own, and in the author's experience with Northern Cheyenne, Navajo, Ute, Wind River Shoshone, Colville, Crow, Southern Paiute, Pueblo, and Chippewa cases, these "mind-sets" have often excluded a consideration or perhaps even a comprehension of Indian cultures.[15] This is not to suggest that employees of many tribal governments do not treat the resources on tribal spaces (including minerals and water) as saleable commodities; but when such tribal resources are evaluated in market terms and especially when they are sold and extraction begins, many native residents disapprove, often bitterly.[16]

Peter MacDonald, Chairman of both the Navajo Tribe and the Council of Energy Resource Tribes (CERT),[17] exemplified the views of some Indians who have striven to sell tribal energy resources and develop tribal economies. He has been dismayed, but not defeated, by the inability of CERT members to develop energy resources to tribal advantage. In early 1976 Chairman MacDonald proclaimed that "we [the Navajo] are going to get our fair share for our energy resources, or we aren't going to share at all....We are giving away the reservation by tons and barrels."[18] Five and one-half years later Chairman MacDonald could cite little real progress toward economic self-sufficiency from energy developments among CERT members. To the contrary, he felt that tribal gains, Navajo in particular, had been set back by the Reagan Administration as public monies had shifted to the wealthy.[19] He recognized a tribal dependency on federal transfer payments even though Indian energy resources were leased and extracted by transnational energy corporations with tribes gaining some lease income, some royalty income, and some jobs. Nevertheless, MacDonald called for greater investments by corporations on Indian reservations, but conceded that development should be consonant with traditional Indian beliefs in the interconnectedness of nature's elements.

Surveys of Indian opinions, however, suggest that traditional concerns for land cannot be accommodated to modern strip mining or in situ extraction techniques, while no known extraction techniques are consonant with Indian views of land.[20] That Indians are destitute, and that energy-related contacts hold the promise of material rewards to alleviate their financial distress, is incontrovertible. The stark conditions, if not dire need of some tribes, account for the contracts that they have entered into. But their lack of information, lack of bargaining power, and their lack of access to finances makes them severely disadvantaged. Corporations have colossal bargaining powers and access to finances and government research; they lease vast deposits of energy resources, exercise immense influences in pricing energy commodities, possess technical advantages over any single tribe or coalition of tribes, and have a wide range of options in energy production and marketing. Indian leaders who wish to lease their tribal resources, therefore, find themselves in tight places. Lease and royalty income helps underwrite a tribal superstructure, providing sound jobs and an expanded tribal budget, but dependency, rather than independent economic growth, has been the outcome.

A number of studies in the High Plains, Southwest, Northwest, and Great Basin have reported favorable Indian attitudes toward energy developments so long as jobs are provided to natives, no changes occur to the environment, no disrespectful white discriminate against Indians, Indian culture is not denigrated, Indian sacred areas are not defiled, and so forth.[21] Even when Indians express desires to obtain jobs, they express skepticism about corporation and federal government actions, are uninformed about who will benefit from development and who will suffer the consequences of underdevelopment and dependency, and they are fearful that their birthright will be transformed.[22] Indian views cannot be understood by trivializing them, while consequences to Indian society from energy developments cannot be understood if they are disregarded. A spate of Environmental Impact Statements, as well as the statements of businessmen, the attitudes of laborers, and the assumptions of social scientists, engineers, and planners have often disregarded Indian views.[23]

Indian Experience with Energy Developments

Except for analytical purposes there is no good reason to separate the business arrangements among tribes, the Department of Interior, and corporations from the consequences to American Indian communities from those business arrangements. I will provide very brief assessments of two cases--Navajo of the Southwest and Northern Cheyenne of the High Plains--before providing a wider set of comparisons. The Navajo began leasing oil and mineral resources in the early 1920s, and the dependency on lease and royalty income in conjunction with federal transfers and welfare payments are deeply embedded features of Navajo society. The Cheyenne present a very interesting contrast.

Navajo. In 1956 Congress passed the Colorado River Storage Project Act. The Glen Canyon Dam on the Colorado River to generate hydroelectric power and the Navajo Dam on the San Juan River (a Colorado R. tributary) were major projects created by the Act.[24]   Both dam projects occurred on Navajo land and the regions and communities adjacent to both experienced boom-bust cycles with long term seasonal development of recreational enterprises. The town of Page, Arizona, adjacent to the Glen Canyon Dam, was built on Navajo land to accommodate construction workers and federal personnel. Between 1957 and 1960, Page grew from zero to 6000, declining to 1200 in 1966. The Navajo Tribe received compensation through land located in Utah, although not along the river. Local Navajos who held sheep permits and browsed their sheep in the area were not compensated for the loss of their land.

Although prior to construction several thousand Navajos comprised the near total population in the 100 square miles around Page, they acquired about 100 of the 1000 construction-phase jobs. Most of the employed Navajos came from outside the local area, all were unskilled, and practically all had left the Page region by 1966.[25]

In 1970 Owens reported that a few local Navajo families remained in Page, few were employed, and of those that were employed, most (84%) were unskilled (laundry or room maids), most were women, and they received low pay (average $2.20 per hour).[26] The better paying jobs, federal jobs included went to non-Indians.

A second boom-bust cycle was precipitated by the first inasmuch as water for coal fired electricity generation was available, infrastructure was developed, and coal was available nearby. The Peabody Coal Company was one of several foresightful energy corporations which had signed coal leases and water agreements with the Hopi and Navajo Tribes through the BIA between 1957 and 1966. Peabody signed leases which gave them the right to strip 65,000 acres of Black Mesa on the Navajo reservation and the Hopi-Navajo joint use area at 25 cents per ton for as long as they continued to extract. The price of the coal was fixed so that no escalator clause operated if the price of coal increased. Coal currently sells at dockside in Los Angeles for over $70 per ton.[27] In 1969, the Navajo Tribe passed a resolution written for it by the Interior Department that limited Navajo rights to Colorado River water to not more than 50,000 acre feet per year of which 34,000 acre feet was to be used, without payment, at the Navajo Generating Station.[28] Peabody also buys 3,000 acre feet of water from the Navajo in perpetuity. The going rate for a consumptive water right (in perpetuity) is between $200 and $2,000 per acre foot in that region, whereas the annual rate (not in perpetuity) should be about $20 per acre foot.[29]

The Navajo Generating Station (a 2,310 megawatt plant), a railroad to transport coal to that plant, a 273 mile slurry line (water driven pipe line) to conduct coal to Bullhead City, Arizona to a 1,580 megawatt plant, and several related projects were begun. Between 1970 and 1976, Page grew from 1,500 to 9,000 residents and the workforce, at the peak in 1973, numbered 2,200 persons.[30]

During the construction phase of the second boom-bust cycle Navajos occupied 20 percent of the jobs. Most of the Navajo employees were in-migrants from other districts on the reservation or from off-reservation locales. The local residents who obtained employment were few, and occupied unskilled jobs almost exclusively, so that there was little transformation of the labor force.

Some employed Navajos, exclusively males, commuted daily from communities 75 miles distant (Tuba City, Kayenta). Many men camped in their trucks and returned home on weekends. Some men stayed in dormitories or even motels in Page. And in a few instances families set up temporary housing near the Lechee Chapter House, or they moved into a trailer camp near Page.[31] The employed Navajos either remitted funds to their homes, or simply returned with them. The jobs payed well, but were temporary. And job income was spent either in off-reservation towns, or in businesses owned by non-Indians. So there was no multiplier effect for the Navajo economy. The profits from coal, water, and electricity, less the small fees, rents, royalties, and one time gifts, accrued to the energy consortia and to Peabody Coal. The Navajos, who were cajolled into giving up their rights to tax the energy companies, observed that the State of Arizona was earning $10.5 million per year from taxes on the Navajo Generating Station in the mid 1970s. That sum was two-thirds the total income of all income (including wages) from all coal and coal-related developments accruing to the Navajo Tribe for the entire reservation. "The Navajo's economy, as measured by personal and tribal income, is steadily losing ground each year in comparison with the United States."[32]

Economic development isn't supposed to work this way, but as has been amply demonstrated in so many cases for American Indians, this is precisely what happens. And just as the Navajos subsidize the U.S. economy through the cheap sales of resources, they are heavily subsidized by state and federal funds. Public funds comprise about three fourths of tribal income.

The economic consequences to Navajos from the Black Mesa-Navajo Generating Station, and from all other coal, uranium, gas, oil, and hydroelectric projects across the reservation, have been only marginally beneficial. The 160,000 Navajos have occupied only 3,000 of the 47,000 jobs that have been available, and those jobs have been temporary, usually restricted to the construction phases of projects.

There has been no internal economic development, and no spin-off employment. Yet non-renewable resources have been drained from the reserve. Great disparities in income have occurred among workers (skilled/non-skilled) and between Navajos (employed/unemployed). Local chapters (vaguely equivalent to town meeting organizations) on the reservation were overwhelmed by the decision made by the energy companies and the central tribal administration to use their water, extract resources, expropriate the graze and browse areas for their livestock, and to cut roads, railroads, and slurries across their terrain. Energy workers moved into some areas seeking better housing and taxing the water supplies.

Deep political divisions occurred between chapters and the central tribal administration. Family based obligations were threatened by migration and the nuclearization of some worker families. Sacred shrine areas were disturbed by energy-related operations, and people's beliefs about their health, the well-being of their families, and all of Navajo culture were threatened.

Since the mid-1970s Navajos have become increasingly skeptical of energy developments as they have become more knowledgeable about them: the absence of information and the desire for jobs made them less skeptical in the 1960s. For instance, one Navajo opinion survey in 1975 demonstrated that those who opposed coal developments outnumbered those who favored them 2 to 1.[33] Another survey conducted by the Shiprock Research Center asked, after mines, railroads, and generating plants were developed in Fruitland (north of Shiprock in New Mexico) and Page (west of Shiprock in Arizona) whether Navajos would support those same developments if they had been forewarned of their probably effects. About 15 percent of all people sampled in the Shiprock District of the reservation gave unqualified "yes" answers, usually because the plants and mines provided jobs. Another 20 percent answered "yes" but with various qualifications, such as "it would be better if Navajos ran the plants." Ten percent said they were not certain, and their responses included "The people who lived on these locations were never asked about the developments," and "There are more Anglos and not enough Navajos working and even if we started over, maybe the same thing would happen." Twenty-five percent were firmly opposed although they varied in their responses. Among this group it was often said that "Mother Earth" should not be dug into, pollution was a desecreation, or that relatives, houses, and stock had been moved. About 30 percent did not know or would not respond.[34]

Navajo people in the same area around Shiprock were surveyed two years later. In this assessment it was demonstrated that many people in the Shiprock area, the center of prospective coal gasification and uranium developments, had no idea what would happen to them if the developments occurred as planned. They did not know how or where they would be relocated, and they expressed two main concerns: the disruption of reciprocity-based kin life and loss of the sheep permit.[35]

Navajo residents claimed that their matri-centered household clusters (several nuclear families of a woman's daughters are clustered near her house) through which land and sheep are inherited would be broken up and the mutual help and assistance patterns that obtain within kinship networks would be disrupted. Moreover, any relocation would not only disrupt reciprocal obligations of food, labor, and ceremonial exchange, but would also be accompanied by the psychological trauma of being removed from lifelong kinspeople. Yet neither the corporations engaged in planning the projects, the BIA, nor the Department of Interior asked the local residents whether they wanted the projects at all.[36]

In a few areas of the reservation where Navajo-corporate employee relations have been sustained over long periods, residents hold very negative attitudes about BIA leasing arrangements and the manner in which lessees treat Navajos. Navajos in the Aneth-Montezuma Creek section of the reservation in southeastern Utah had been irritated by gas and oil operations conducted by Texaco, Phillips, and other lessees for over two decades. Few Navajos were hired to work on the oil operations, and non-Navajo employees of the oil companies frequently ridiculed and mistreated local Navajos. Furthermore, Navajos were angered at the lease arrangements worked out by the BIA in the late 1950s. In April 1978, Navajos in the Aneth region rebelled and took over all the gas and oil operations, driving off the workers and shutting them down completely. They demanded that the oil leases be voided, or renegotiated, that Navajos be hired for jobs, that the oil companies make substantial financial contributions to Navajo educations, and that discriminatory acts by non-Navajo employees of the oil companies cease immediately. After two weeks the oil companies complied with all demands except that which required renegotiation of the leases let by the BIA in the 1950s. For so long as the companies continued to produce, the leases were unchallengeable.

Interestingly, the Aneth area Navajos did not vent all of their displeasure on the corporations and the BIA. The Aneth residents also castigated the Navajo Tribal Council for taking the money obtained from oil leases and royalties at Aneth, but providing few resources and services to the Aneth people in return.[37] Aneth Navajos demanded more resources and services from the tribe, demanded to know what transpired at the seat of Navajo government, and demanded that the corporations settle with the local residents rather than the tribal government in Window Rock. The centralized government drains off the money for its own ends, but provides little in return, and the locals in the Aneth chapter area resent it.[38]

At the time of first contact with Europeans the Navajos were not organized as a tribe. That is to say, the Navajos recognized no chief or council of chiefs with the authority to govern all Navajos and to make binding decisions of any type for all Navajos. Tribal organization is solely an outgrowth of federal government relations with the Navajos in the twentieth century. Indeed, the Navajo Tribal Council was formed by the Department of Interior in 1923 so that the Department could negotiate oil leases. The large numbers of Navajos that are administered by the tribal government are seldom consulted by tribal decision makers, just as during the course of Navajo relations with industries and the federal government, the Navajo tribal government has seldom been consulted or fully informed by these bodies.[39]

The economic, social, political, and religious consequences to Navajos from energy developments have been considerable, but seldom beneficial. Benefits have accrued to corporations, to energy users in California and the Southwest, and to the Navajo Tribal Government, which has grown in size and accumulated experience and expertise, but which has also become dependent on very large sums from public funds and lease and royalty income.

Northern Cheyenne. The Northern Cheyennes of southeastern Montana present a very different culture history from that of the Navajo. Prior to reservation subjugation, the Cheyenne possessed a tribal government directed by a Council of 44 Chiefs and protected by the actions of a Council of 24 Chiefs to administer tribal affairs. Although these native tribal bodies withered in the nineteenth century following the expropriation of Cheyenne resources and the incarceration of Cheyennes at several forts managed by the United States Cavalry, the sentiments of tribal organization and

traditional Cheyenne culture have been resilient. The
Northern Cheyenne and their Crow neighbors in Montana reside
upon perhaps more than ten billion tons of extractable coal.
The manner in which the Cheyennes have deliberated about what
to do with their resources is most instructive.

In the 1960s and 1970s, the Northern Cheyennes, similar
to their energy-rich Indian neighbors in the West, were pulled
by coal extracting corporations with the lure of fixed royalties to yield the coal beneath their homes and homeland.
Economic consultants, the BIA, and the Interior Office encouraged and advised the Cheyenne to sell. The tribal government and the tribal elders, beginning in 1973, resisted, pre-
ferring to establish their own course of action consonant
with Cheyenne traditions.

Less than a decade ago, in 1973, pursuant to a discussion
with a coal company about signing a lease to allow for the
extraction of coal from the reservation, the Northern Cheyenne tribal council learned that fully 56 percent of the
reservation's land surface had already been leased to coal
companies by the BIA for extraction, and that the lease
royalties they were to receive were fixed at 17.5 cents per
ton. That is, as long as the companies continued to extract
coal, and no matter what price a ton of coal brought on the
market, the Northern Cheyenne would receive 17.5 cents per
ton.[40] No coal companies had begun extraction at that time,
and no company has been able to begin operation since. In
1973, when the Northern Cheyenne realized that they had
signed such unfavorable leases (36 violations of federal
leasing producers alone were discovered), Northern Cheyenne
families were paying $26 per ton for coal hauled in from
Sheridan, Wyoming, a few miles south of the reservation.

The Northern Cheyenne tribal council, urged on the one
hand by their elders to respect and protect the wholeness
and interrelatedness of their land, air, and water, and
angered on the other hand by the incomplete and misleading
information given to them by the BIA and the corporations
when they signed the leases, sought to revoke all coal contracts. In 1974 the tribe retained the Seattle law firm of
Ziontz, Pirtle, Morisett, and Ernstoff, to petition Secretary
of Interior Rogers C. B. Morton to cancel the leases.
Because of the legal violations in the lease agreements and
the federal government's failure to dispatch its trust obligations to the Northern Cheyenne, Secretary Morton suspended
but did not cancel the leases. Over the next seven years the
tribe expended $200,000 in legal fees to get the leases can-
celled outright. The Cheyenne could seek cancellation because they had not become dependent on the lease income and
extracting royalties.

As the Cheyennes were apprised of the probable consequences to their reservation that would result from the
stripping of Cheyenne coal and the prospects of mine mouth
power plants to burn it, they also became more completely
aware of the probable consequences to the Cheyenne reservation from other electricity generating projects in the adjacent region. Only 15 miles to the north of the reservation,
at Colstrip, two coal-fired electricity generating plants
were being readied for operation, and there were plans to add
two more plants to the first two. At Decker, hard by the
southern border of the reservation, the largest open pit coal
mine in the world was in operation. Fully 47 separate mines,
electricity generating plants, and coal gasification plants
were either in operation or were scheduled for operation
around the perimeter of the reservation.

In response to the prospect of the effects that such a
combination of industrial operations posed for Cheyenne life,
the Northern Cheyennes, through their own Northern Cheyenne
Research Project which the Tribe created to investigate the
consequences from energy developments and to analyze requests
for their resources, applied to the Environmental Protection
Agency to change the air quality status of the reservation
from Class II to Class I. A report and case was prepared
and submitted. The request was granted, assuring the Northern
Cheyenne that there would be no significant deterioration of
their air quality. Thus, local industrial facilities were
not allowed to pollute Cheyenne air, and further developments
at Colstrip were stopped.

So after 1973, the Northern Cheyenne tribe began to gain
for itself and through its own auspices the information that
it required so as to make informed decisions about its economic and industrial destiny, about its quality of life, and
about how the reservation would be preserved and used. The
Northern Cheyenne Project, headed by Cheyennes, set about to
collect and analyze data relevant to the Cheyenne reservation
and Cheyenne society. When pushed toward economic projects,
the Northern Cheyennes demanded information. The tribe opened
discussions with non-Indian ranchers in the vicinity who were
battling energy corporations, public utilities, and the
federal government to protect their land and their stock-
raising way of life from coal shovels, noxious air, and
acidic rain.[41] Though non-Indian ranchers, like white residents in towns surrounding the reservations of southeastern
Montana, had discriminated against Indians for generations,
the old contempt was lessened by the new alliances and
the cause that the two populations and the two cultures
shared.[42]

Cheyennes freely discussed the difficulties with which their ancestors coped over the past 200 years--battles and agonies that living Cheyennes refuse to forget. The Northern Cheyenne Reservation, in their view, was bought with the lives of Cheyenne people for time immemorial, and modern Cheyennes are obligated to preserve it by looking out for the interests of the elderly, the young, and future generations.[43]

The current Cheyenne tribal government was created following the Indian Reorganization Act of 1934. Though the elderly Cheyennes and even younger traditionals have little to do with the operations of the tribal government--often the two are in conflict--the wisdom of the elders has served to counsel the tribal decision makers. The conservative messages of elders harken to the precepts of the tribe, and on key issues, such as coal extraction, the message spreads through families and between generations as people seek understanding from elders in the major Cheyenne rituals, such as Arrow Renewal, Medicine Hat Renewal, Sun Dance, and Buffalo Men, through Peyote ceremonies, and in everyday affairs of life.

As the tribal council (tribal government) sought to make the tribal economy viable while overseeing the political and legal affairs of the tribe, it also sought information to help evaluate the claims of corporations that desired to extract Cheyenne coal and water. And it asked the Cheyenne people what kind of developments they preferred and what they wanted to maintain from their present lives.[44]

Cheyennes responded overwhelmingly with expressions of desire to maintain their language, religion, and old ways for future generations, and they sought solutions to alcohol abuse, inadequate social services, and unemployment. At the height of employment activity during the construction of the coal-fired plants at Colstrip in 1974, only 34 Cheyennes could secure employment in the work force of 895 people, and the number eventually dropped to six employed Cheyennes in 1976.

The experiences at Colstrip, and the experiences of Anglo residents in rural southeastern Montana towns that have sustained rapid growth through industrial energy developments, have not been lost on the Cheyenne. Few jobs are generated for local residents; economic controls and ownership reside with corporations outside the local areas; electricity is delivered to businesses and homes in distant cities; resources are drained from the local area; immigrants do not understand or respect local traditions; public services are strained beyond the breaking point; community frictions increase; and the environment is degraded.

Cheyennes feared all of these consequences and many more.
When asked what kind of business developments they desired,
Cheyennes suggested residentiary businesses so that they did
not have to drive to Billings and other distant places for
durables, and they suggested recreation projects and local
theaters so as to entertain themselves and allow outsiders to
enjoy some of the beauty of their reservation. But although
they desired local businesses so that Cheyennes could shop
conveniently and keep their money circulating locally, they
also felt that the businesses should be tribal operations for
the good of all and not for the benefit of a few. Livestock
and farming operations were high on the list as well; but
coal ranked far down on the list of businesses that Cheyennes
desired to develop. Moreover, upon further questioning it was
learned that Cheyennes wanted a small coal-mining operation
to provide for local Cheyenne needs, not for the monentary
energy desires of Chicagoans, or the profits of Westmoreland
Coal.

When asked about large coal extraction schemes for the
reservation, Cheyenne residents felt that positive benefits,
such as money, jobs, availability of coal for use by Cheyennes, and increased tribal control over tribal resources,
were outweighed 2 to 1 by negative effects, especially the
anticipation of increased social and community problems,
damage to the environment and the loss of non-renewable resources, and the influx of whites who would have no respect
for Cheyenne culture.

Cheyennes frequently referred to Colstrip as an example
of the destruction of a small town. Cheyennes did not want
the same course of events to overtake them. They felt that
they suffer discrimination away from home and did not want
to suffer it at home as well. They sought to preserve what
they had. Eighty-seven percent of Cheyennes wanted no whites
on the reserve. They did not want the land abused, yet they
were willing to manage and extract renewable resources.
Cheyennes repeatedly said the land is their home and they
treat it as a person.

On October 31, 1980, Congress passed special legislation
for the Northern Cheyenne, PL 96-401, that allowed Secretary
of Interior Cecil Andrus to cancel one of the many lease
contracts between the Northern Cheyenne and a coal company.
This one, with Peabody, granted them rights to about one
billion tons of coal.[45] Peabody, as compensation, received
the right to lease, non-competitively, 11,000 federal acres
adjoining other Peabody lease tracts in Montana. Their lease
fees were deducted from the fees owed to the federal govern-

ment, as were all investments that Peabody had made on the
reservation.

Yet during the year prior to the cancellation of the
Peabody contract, the tribe was under increasing financial
pressure as it had engaged in several expensive legal battles
(1) to get Congress to enact legislation that would allow for
the cancellation of contracts, (2) to exercise its water
rights over State opposition, and (3) to stop Montana Power
Company from developing Colstrip generating units 3 and 4.
According to Boggs[46] "The feeling was that the tribe was
running out of economic and political options." He cited
Allan Rowland, tribal chairman as saying "we've made million-
aires out of several lawyers....We blocked Colstrip 3 and 4
for seven years before we ran out of money. If someone would
give us the money we'd block the Montco Mine and TR Railroad."

Between about September, 1979 and May, 1980 tribal displeasure grew over the financial crisis, and factions within
the tribal government formed over the issue whether or not
the tribe should sign a contract with ARCO to explore for oil
and gas. In May, 1980 the group led by a Harvard Business
School-educated Cheyenne favoring the contract, which was
written by ARCO, prevailed. The opposition, whose argument
was based on Cheyenne cultural values and an unwillingness
to see the Northern Cheyenne position on natural resources
and sovereignty reversed, said the agreement was "ramrodded
through."[47]

The opposition went to the community and gained 400 sig-
natures for a referendum against the agreement. The Council
voted against a referendum, whereupon the Tribal Judge issued
an injunction against the agreement (she was fired immediately). The Council held a referendum in which the agreement
was approved, and $1,500, which had been promised to every
Northern Cheyenne contingent upon approval of the agreement,
was paid from a $6 million up front gift from ARCO.

The tribal council was under intense economic pressure,
and oil and gas exploration seemed much less dangerous than
coal extraction. The tribal members were uninformed about
the details of the agreement, which had been written by ARCO.
The agreement provided $6 million in a signature bonus and
25 percent production share to the Northern Cheyenne. It
also gave the tribe little control over information generated
on its lands, and provided a 20 year tax holiday, and free
access to tribal lands, roads, and water, to ARCO. The agreement fails to provide a role for the tribe in planning oil and
gas exploration or development on the reservation.

General Conditions of Economic Contracts with Indian Tribes

Whether in leases, profit-sharing schemes, or joint-ventures with transnational corporations, Indian tribes lack capital, power and expertise, and usually have been misinformed and il-advised. With one exception, Indian tribes that own energy resources do not control production, sell the products, or keep the books on any reservation. The exception is the Jicarilla Apaches of northern New Mexico, who assumed ownership and control of six oil and gas wells from its joint venture partner, Palmer Oil, in July 1980.

The general condition is very different, and the following is a tiny portion of that condition. The Jicarilla Apaches succeeded in opening the books of an oil corporation that was extracting Apache-owned oil and found a $600,000 payment deficit. In December, 1980, the Wind River Shoshones and Arapahos of Wyoming alleged that about $3 billion worth of their oil had been stolen by companies and persons engaged in extracting, storing, and transporting the oil. Indictments have been brought down, and a Department of Interior investigation concluded in early 1982 concurred with many of the allegations brought by the tribe. Litigation will follow.

In recent years the Navajo Tribe has attempted to move from lease and royalty arrangements to joint ventures. In 1975 the Navajo Tribe signed a joint-venture agreement to explore for, and to extract uranium with EXXON, in which EXXON provided three options to the tribe. Crucial to all three options was the unstated premise that the very uranium that EXXON wished to extract had no value.[48] Moreover, EXXON would keep the books, sell the product, etc., and for every dollar that EXXON provided toward the project on behalf of the tribe, EXXON would be paid back double their contribution and 177 percent flat profit.

In 1975 a group of 25 energy resource rich tribes, following the advice of business leaders, government leaders, and a former government employee who became the director of the organization that he encouraged them to create, formed the Council of Energy Resource Tribes (CERT). The member tribes receive information from CERT about lease royalty agreements, and CERT lobbies congress and federal agencies for funds and benefits. CERT has also sought to bring the energy corporations and tribes together. CERT did so in 1980 when the Western Regional Council (WRC), comprising chief executive officers of the largest energy-related corporations, including utilities and banks in the Western states, met in Phoenix at the same time as the annual CERT meeting, but at a different hotel. WRC sponsored cocktail

parties for CERT members, realizing that in order to survive
they can no longer, as one WRC executive put it "treat these
situations [dealing with Indians through leases, joint ventures, or other contracts, so as to acquire their energy re-
sources] as a 100-percent profit orientation because we also
need to develop our reputations...[and] they [the Indians]
have to feel they made a good deal."[49]

In 1981 CERT invited businessmen to its meeting in
Denver, the theme being "Doing Business with Indian Tribes."
The transnational corporations, including WRC members, were
well represented, and they heard chairman MacDonald say that
the energy corporations and the companies can be allied, but
companies will have to consider a return of less than 40 percent from investments on Indian reservations to do so.

Corporations possess capital, information, and access to
power. The federal government provides research and conducts
business for tribes and corporations. Indians possess little
capital, little information, and scant access to power.
Energy tribes have become dependent on federal largesse and
royalty payments, and these dependencies cause them to get
pinched when payments from either source dwindles. The pinch
is felt in households where basic necessities are not adequately met. Tribes often enter into contracts, uninformed,
so as to maintain the tribe's job structure, keep cash flowing through it, and plan for self-sufficiency. So far, self-
sufficiency has been a chimera. The list of complaints that
have stemmed from most decisions to allow external corporations to develop energy resources on reservations is long.
Families have been relocated, lost their grazing permits,
and lost their houses. They have lost their attachments to
the places in which they have reared their children and which
they expected to pass on to their children's children. Extended family households have been severed. Burial grounds,
shrines, and sacred areas have been desecrated, or threatened
with desecration. Disputes within tribes have intensified
and intense factional disputes have emerged. To top it off,
jobs, except for the most menial types of general labor and
custodial maintenance, have gone to in-migrants.

It is not appropriate to consider tallying up costs and
benefits for American Indians from energy developments. That
task should be left to scholars who trivialize or disregard
Indians and their views. Indians do not control energy developments, they respond to them, often in desperation, gaining some lease income as their non-renewable resources are
drained off their reserves.

Well-meaning persons have often encouraged developments

whose negative impacts went far beyond those that had been anticipated. Part of the problem is the very nature of profit maximizing in capitalist enterprises, part is in the domination of tribes by the federal government and tribal dependencies on public funds, and part is that we Anglos continue to forget how culture-bound we are. Perhaps the clearest lesson is that the approaches of the past haven't worked, and that we're going to need new ways of thinking about energy development on or near Indian lands in the future.

REFERENCES

1. "Anglo" is commonly used in the social science of the west to distinguish non-Indians and non-Mexican-Americans from Indians (Native Americans) and Mexican-Americans.

2. See the following for accounts of recent interest in Indian energy resources. Joseph G. Jorgensen et al., 1978, Native Americans and Energy Development (Anthropology Resource Center: Cambridge); Joseph G. Jorgensen, 1982, Native Americans and Energy Development II (Anthropology Resource Center and Akwesasne Notes: Boston and Owlshead); Roxanne Dunbar Ortiz, editor, 1980, American Indian Energy Resources and Development (Institute for Native American Development, University of New Mexico Development Series No.2).

3. See especially Lee Brown, 1981, "Conflicting Claims to Southwestern Water: The Equity and Management Issues," The Southwestern Review of Management and Economics I (1): 35-60; and Noel R. Gollehon et al., 1981, "Impacts on Irrigated Agriculture from Energy Development in the Rocky Mountain Region," The Southwestern Review of Management and Economics I (1): 61-88.

4. Public Law No. 383-73D Congress.

5. See the provisions of the Act and see the discussion of "domestic, dependent nations" in Joseph G. Jorgensen, with Shelton Davis and Robert Mathews, 1978, "Energy, Agriculture, and Social Science in the American West" In Jorgensen, et al., Native Americans and Energy Development, for an assessment of the limited powers vested in tribes and the extensive powers over Indian tribes vested in the Secretary of Interior and committees of Congress.

6. See, for example, Business Week, 27 January 1975, p. 108; Business Week, 3 May 1976, pp. 100-102; Los Angeles Times 10 July 1977, Part I, p. 20; Los Angeles Times 16 October 1977, Part I, pp. 3, 30-32; Los Angeles Times 19

October 1980, Part I, pp. 1, 12-13; Lynn Arnold Robbins, 1979, "Navajo Energy Politics," The Social Sciences Journal 16(2): 93-119; Council of Energy Resource Tribe's slide show and movie, narrated by Lorne Greene and viewed nationally in late 1981; Bruce M. Rockwell (Chairman, Colorado National Bank), 1981, speech delivered in Denver and published nationally by Associated Press; Dr. Michael Halbouty (consulting Geologist and Engineer), 1981, speech to Western Regional Council in Denver; Bob Gottlieb and Peter Wiley Straight Creek Journal, 20 March 1980, pp. 1-3.

7. See, for example, the Navajo Tribe-EXXON contract to explore for uranium and its aftermath in Joseph G. Jorgensen, 1978, "A Century of Political Economic Effects on American Indian Society," The Journal of Ethnic Studies 6(3):53-56; the Burnham District coal-gasification project in Robbins "Navajo Energy Politics;" the Dalton Pass uranium project in Joseph G. Jorgensen, 1982, "The Political Economy of the Native American Energy Business" In Jorgensen Native Americans and Energy Development II: the Northern Cheyenne response to oil and gas issues signed by the tribe in James P. Boggs, 1982, "The Challenge of Reservation Resource Development: A Northern Cheyenne Instance," In Jorgensen, Native Americans and Energy Development II; the Colville response to a tribal contract with AMAX in Jean A. Maxwell, 1982, "Colvilles on the Verge of Development," In Jorgensen, Native Americans and Energy Development II; the traditionalist Hopi villages responses to Hopi IRA government complicity in the contracts with Peabody Coal for Black Mesa extraction in Richard O. Clemmer, 1978, "Black Mesa and the Hopi" In Jorgensen, Native Americans and Energy Development, pp. 17-34.

8. See especially the thirteen explicit demands and the physical responses of Aneth area and Montezuma Creek area Navajos to Texaco, Conoco, Phillips, and Superior Oil operations in their areas in Tom Barry, Navajo Times 13 April 1978; Gaylord Shaw Los Angeles Times 13 April 1978, Part I, pp. 1, 21.

9. See Thayer Scudder, 1979, Expected Impacts of Compulsory Relocation on Navajos with Special Emphasis on Relocation from the Joint Use Area (Institute for Development Anthropology: Binghamton); Newsweek, 9 April 1979, p. 98.

10. The Rand Corporation study by Richard Nehring et al., 1976, Coal Development and Government Regulation in the Northern Great Plains (Rand Corporation: Santa Monica) points out that in deciding to participate in an energy extraction-conversion project "firm and industry behavior is characterized by standard profit maximization....Corporations and the

consumers they serve may realize all the benefits...without having to bear all of its costs."

11. See Laura Nader, 1979, "Barriers to Thinking New About Energy" (Mitre Corporation: McLean, VA); Laura Nader, et al., 1980, "Supporting Paper 7, Energy Crisis in a Democratic Society," Study of Nuclear and Alternative Energy Systems (National Academy of Sciences: Washington, D.C.); Joseph G. Jorgensen, 1980, "Social Scientists vs. the National Academy of Sciences' Mining Lobby" (unpub. ms.); Jorgensen, "The Political Economy of the Native American Energy Business."

12. For comparative analyses of American Indian societies see Alfred L. Kroeber, 1939, Cultural and Natural Areas of Native North America (University of California Publications in American Archaeology and Ethnology 38: Berkeley); Harold E. Driver and William C. Massey, 1957, Comparative Studies of North American Indians (Transactions of the American Philosophical Society 47(2): Philadelphia); Joseph G. Jorgensen, 1980, Western Indians (W.H. Freeman and Co.: San Francisco).

13. See statements from Hopi and Navajo leaders in Diane Weathers "A Fight for Rites," Newsweek 9 April 1979, p. 98; statement in testimony of Russell Jim, Yakima councilman, Subcommittee on Nuclear Regulation, Senate Committee on Environment and Public Work, 24 January 1980; statement in petition of Navajo residents of Dalton Pass in seeking to halt all uranium mining in their region "Wherever There is Energy Development, There also seems to be 'Grass Roots' Opposition" Navajo Times 15 June 1978; Plaintiff's Brief, Starlie Lomayaktewa et al. v. Rogers C. B. Morton and Peabody Coal Co. (D.D.C. 14 May 1971) Exhibit A, p. 1.

14. See, for examples, Lynn A. Robbins with Robert Stewart, 1979, "The Socioeconomic Impacts of the Proposed Skagit Nuclear Power Plant on the Skagit System Cooperative Tribes (Lord and Associates: Bellingham), p. 45, for survey results on the Skagit and Swinomish Tribes; G. Mark Schoepfle et al., 1978, A Study of Navajo Perception of the Impact of Environmental Changes Relating to Energy Resources Development (Navajo Community College: Shiprock); Jean Maxwell Nordstrom et al., 1977, The Northern Cheyenne Tribe and Energy Development in Southeastern Montana, Vol. I: Social, Cultural and Economic Investigations (Northern Cheyenne Research Project: Lame Deer).

15. When reporting to a Committee of the National Academy of Sciences in 1978 about Indian concepts of ownership, stewardship, and economic distribution, an economist asked this writer if I could explain how it was that western

economists came to take, as given, the concept of economic
man.

16. Several Navajo, Colville, Northern Cheyenne, and
Hopi instances have been cited above. The bitterness, and
tears, have accompanied many Indians to federal courtrooms.

17. The Council of Energy Resource Tribes, formed
through the impetus of corporation advisors and disgruntled
Indian leaders, comprises 25 member tribes--mostly in the
West and mostly the possessors of abundant energy reserves.

18. "Indians want a Bigger Share of their Wealth,"
Newsweek, 3 May 1976, pp. 100-102.

19. Peter MacDonald, keynote speech "Doing Business
with Indian Tribes," CERT Annual Meeting, October 1981
(Denver).

20. See Committee on Soil as a Resource in Relation
to Surface Mining for Coal, 1981, Surface Mining: Soil,
Coal, and Society (National Academy Press: Washington,
D.C.), esp. pp. 31-35, for statement about frictions between
pro-development and more traditional Indians who resist the
alteration of the natural landscape. The natural landscape
can be reclaimed, but not restored, and there is no currency
to trade for the lost symbolic meanings of land.

21. See, for examples, Robbins and Stewart, The Socio-
economic Impacts of the Proposed Skagit Nuclear Power
Plant...; Nordstrom et al., The Northern Cheyenne Tribe and
Energy Development in Southeastern Montana; Harris Arthur and
Barbara West, 1976, "Preliminary Survey Results: Attitudes
Towards Existing Development," (Shiprock Research Center:
Shiprock); Crow Impact Study Office, 1977, A Social, Economic
and Cultural Study of the Crow Reservation: Implications for
Energy Development (Crow Agency, Montana).

22. A sampling of the tribes that have experienced all
or most of these consequences and fears are (1) Hopi
(Richard O. Clemmer, 1978, "Black Mesa and Hopi" In Jorgensen
et al., Native Americans and Energy Development, pp. 17-34,
and 1982, "Effects of the Energy Economy on Pueblo Peoples"
In Jorgensen et al., Native Americans and Energy Develop-
ment II); (2) Navajo (Lynn A. Robbins, 1978, "Energy Develop-
ments and the Navajo Nation" In Jorgensen et al., Native
Americans and Energy Development, pp. 35-48, and "Energy
Developments and the Navajo Nation II" In Jorgensen et al.,
Native Americans and Energy Development II; G. Mark
Schoepfle et al., 1981, A Study of Navajo Perceptions of the

*Impact of Environmental Changes Resulting from Energy Re-
source Development*, Report submitted to the U.S. Environ-
mental Protection Agency by Navajo Community College: Ship-
rock, New Mexico); (3) Cheyenne (Nordstrom, 1977, *The
Northern Cheyenne and Energy Developments in Southeastern
Montana. Volume I: Social, Cultural and Economic Investi-
gations*; James P. Boggs, 1982, "The Challenge of Reservation
Resource Development: A Northern Cheyenne Instance" In
Jorgensen et al., *Native Americans and Energy Development
II*); (4) laguna, Acoma, Eastern Keresan, and Tanoan Pueblos
(Richard O. Clemmer, "Effects of the Energy Economy on
Pueblo Peoples)); (5) Colville (Jean A. Maxwell, 1982,
"Colvilles on the Verge of Development" In Jorgensen et al.,
Native Americans and Energy Development II); and (6) Crow
(Jean A. Maxwell and the Crow Impact Study Office, 1977,
*A Social, Economic and Cultural Study of the Crow Reservation:
Implications of Energy Development*. 2 vols. Montana Crow
Tribe). (7) Skagit and Swinomish (Robbins with Stewart,
*The Socioeconomic Impacts of the Proposed Skagit Nuclear
Power Plant...*).

23. For examples see Tennessee Valley Authority and
Department of the Interior, April 1978, *Final Environmental
Statement Dalton Pass Uranium Mine*; Walter H. Peshlakai *et
al.* vs. James R. Schlesinger *et al.*, U.S.D.C. for the Dis-
trict of Columbia, Civil No. 78-2416; Jorgensen, "The
Political Economy of the Native American Energy Business,"
Joseph G. Jorgensen, 1981, "Social Impact Assessments and
Energy Developments," *Policy Studies Review* I (1); James P.
Boggs, 1978, "Relationships Between Indian Tribes, Science
and Government in Preparing Environmental Impact Statements"
Northern Cheyenne Research Project Paper Series (Northern
Cheyenne Research Project: Lame Deer); Robbins with
Stewart, *The Socioeconomic Impacts of the Proposed Skagit
River Nuclear Plant...*; Gottlieb and Wiley, "The New Power
Brokers who are carving up the west."

24. See, for a sampling of the research literature on
the topic of Navajos and energy developments, David F.
Aberle, 1969, "A Plan for Navajo Economic Development," in
Development Prospects and Problems (vol. I, part I) *Toward
Economic Development for Native American Communities*, U.S.
91st Congress, 1st session, Joint Economic Committee, Subcommittee on Economy in Government (Washington, D.C.: U.S.
Government Printing Office); Lynn A. Robbins, "Energy Developments and the Navajo Nation;" Nancy J. Owens, 1979,
"The Effects of Reservation Bordertowns and Energy Exploita-
tion on American Indian Economic Development" *Research in
Economic Anthropology* 2; Donald G. Callaway, Jerrold E. Levy,
and Eric G. Henderson, 1976, "The Effects of Power Production

and Strip Mining on Local Navajo Populations," <u>Lake Powell Research Bulletin</u>, NO-22 (Institute of Geographics and Planetary Physics: University of California, Los Angeles).

25. Donald Callaway, Jerrold R. Levy, and Eric R. Henderson, "The Effects of Power Production and Strip Mining on Local Navajo Populations."

26. Owens, "The Effects of Reservation Bordertowns and Energy Exploitation on American Indian Economic Development."

27. Jorgensen, "The Political Economy of the Native American Energy Business."

28. The consortium that operates the station made a gift of $125,000 to the Navajo Community College and agreed to hire some Navajo employees. The 34,000 acre feet of water has a minimal non-perpetuity value in 1980 dollars of $680,000 annually. The perpetuity rights are valued between $6.8 and $680 million.

29. See Brown, "Conflicting Claims to Southwestern Water," and Gollehon, "Impacts on Irrigated Agriculture from Energy Development in the Rocky Mountain Region."

30. Callaway, Levy, and Henderson, "The Effects of Power Production and Strip Mining on Local Navajo Populations."

31. Callaway, Levy, Henderson, The Effects of Power Production and Strip Mining on Local Navajo Populations," p. 11.

32. Robbins, "Energy Developments and the Navajo Nation," p. 39.

33. <u>Business Week</u>, 3 May 1976, p. 102.

34. <u>Shiprock Research Center Newsletter</u>, March 1976, pp. 1-10.

35. Schoepfle, <u>et al.</u>, <u>A Study of Navajo Perception of the Impact of Environmental Changes Resulting from Energy Resource Development</u>.

36. See the several affidavits of Joseph G. Jorgensen, Lynn A. Robbins, and Ronald L. Little on the WESCO and EXXON-BIA uranium projects.

37. Tom Barry, "Aneth people tired of exploitation: oil wealth has not brought them a better life." Navajo Times, 13, April 1978, pp. A2-3, 6.

38. In several places the author has explained these relations as "metropolis-satellite" (see Jorgensen "A Century of Political Economic Effects on American Indian Society").

39. See Aberle "A Plan for Economic Development;" Robbins, "Navajo Energy Politics," "Energy Developments and the Navajo Nation," "Energy Developments and the Navajo Nation II."

40. Nordstrom, et al., Social, Cultural and Economic Investigations. The Northern Cheyenne Tribe and Energy Developments in Southeastern Montana, 1977, and Nancy Owens, "Can Tribes Control Energy Developments," (In Joseph G. Jorgensen, et al., Native Americans and Energy Developments, 1978, pp. 49-62) provide much of the data on which the Northern Cheyenne example is based.

41. Raymond Gold, 1978, Social Impacts of Strip Mining and other Industrializations of Coal Resources (Institute for Social Science Research: Missoula).

42. Raymond Gold, 1979, "Energy Development-Related Changes in Social and Cultural Patterns in Contemporary Northern Plains Boomtowns" (University of Montana: Missoula).

43. Nordstrom, et al., The Northern Cheyenne Tribe and Energy Development in Southeastern Montana.

44. Several reports and papers from the Northern Cheyenne Research project, but especially Nordstrom, et al., Social, Cultural, and Economic Investigations..., provide results from these investigations.

45. At 1980 prices Peabody could have sold the coal for about $71 billion in Los Angeles (less the tax deductible costs of transportation, lease royalities, some large shovels, some labor time, and the like). The Cheyenne share would be $175 million.

46. Boggs, "The Challenge of Reservation Resource Development: A Northern Cheyenne Instance."

47. Boggs, "The Challenge of Reservation Resource Development: A Northern Cheyenne Instance."

48. That is to say, the Navajo received no capital credit for their resource reserve. Put in other words, and by analogy, it is likely that had EXXON owned the reserves and had the Navajo Tribe sought to do business with EXXON, that EXXON would have accepted the Navajos as a limited partner entitled to a profit percentage less than their percentage contribution to this investment, because the reserves owned by EXXON possessed capital value.

49. Gottlieb and Wiley, "New Power Brokers and Carving up the West," pp. 1-3.

Allen V. Kneese

Closing Comments: Resources and the Environment

The paradox of resources development in the semi-arid to arid West surrounds the conflict between resources development and the environment. At issue is an arid mountainous region that is ecologically, economically, and culturally delicate. The region presents a pattern of contrasts, conflicts, and opportunities. On the one hand, the amenities of the region--clean air, expansive vistas, sunshine, mountain wilderness, cultural diversity--are a great attraction for people and industry. On the other hand, potential exploitation of natural resources, especially energy resources, threatens to degrade and perhaps destroy many of these amenities. The problem faced by the region, and the nation, is how to craft policies and implement procedures that will minimize these conflicts.

Each chapter in this book provides useful background information. Some of the earlier chapters are especially helpful in explicating the geological and technical dimensions of the problem. Others address questions about existing environmental policies. In my commentary, however, I concentrate on the topics of the later chapters, starting with water resources. I do this partly because I think these issues are the more interesting ones and partly because they lie closest to my own expertise.

With respect to water and energy development for the next few decades, energy development may present a substantial challenge to existing institutions, but water availability as such is not likely to constrain the overall level of development. The pattern of development may, however, be considerably affected by laws and institutions, for example, laws that prohibit the transfer of water across the state lines.

Maintaining air quality is likely to present more difficult problems than is water availability. Regional haze could endanger the landscapes in the West to which people assign large values. The air quality studies indicate that tight control of air pollution to safeguard visibility is an economically justified proposition in the area. My question is whether air pollution will be controlled or whether, because of inadequate policies or implementation, we must suffer those environmental damages if development occurs.

The sociological presentations concerned two issues I wish to discuss. One is the importance of bringing sociological concepts into impact analysis. The other is a concern with the accuracy of projections that are made for impact analysis. Firstly, I think there are ideas and insights presented that could result in workable improvements in performing analyses. For example, I was particularly struck by the idea of invisible arrangements in one presentation. Never having thought much at all in sociological terms, I considered that idea important. And it is, simply, that in stable communities there are many informal relationships through which people help and protect each other and that those relationships may be lost in situations of rapid growth. At the same time, as there are more people to help and protect, less assistance happens spontaneously, and more responsibility is shifted to the public sector. That seems to me not only an important idea but a researchable one. Such research would require that resources be devoted to monitoring and understanding the anatomy, so to speak, of development.

In evaluating projections for impact analysis, I particularly liked the effort to make a retrospective study of how past projections have performed. In economic analysis, there are very few retrospective studies. Projections are made, plans are made, benefit-cost analyses are done, but rarely does anyone look back to see whether, in fact, anything like what was projected happened. So I am happy to see that effort being made although it may be still rather rudimentary. Some further discussion may be appropriate.

From a methodological point of view, one might be interested more in whether the projections were accurate in the rate of predicted change than in the absolute number of persons projected for some target year. Errors in rate of change will cause huge divergencies in absolute numbers over, say, the forty- or fifty-year projection periods that one often encounters. In addition, there is the question of valid data. Problems in the projections occasionally might stem from inaccurate census data. Some of the areas that are being projected for are remote, rural, and poverty-striken. Variation from one census to another of persons included and persons excluded could be a source of inaccuracy.

Finally, in situations with a small economy, a small population, and a potentially large single investment, the timing of a large natural resources investment becomes critically important in determining what the projection will show for a given year. One should emphasize, both in terms of economics and in terms of demography, that developed and undeveloped situations are completely unlike one another. The already economically developed situations have multiple forces producing further development. The undeveloped situations are anecdotal in a sense; no underlying pattern of regular change is present. The investment is a large insertion into a small system. Consequently, it might be useful to try to look behind some of those projections to see whether some of them were wrong simply because of an unexpected delay in a large project. This could have a relatively huge impact in some instances, and most projects have unexpected delays. So the retrospective study of projections appears to be an excellent area to explore and appears to merit further work, but present efforts still seem rudimentary.

Another methodological comment about projections is in order. One of the great weaknesses of environmental impact statements like those analyzed in one of the sociological papers is that they take a one-at-a-time approach to each project. Such statements analyze one gasification facility, one power plant, or one other facility at a time. From an environmental point of view, a population-distribution point of view, and a resource-impact point of view, one project may not matter much; however, looking at the whole system of developments occurring throughout a region may produce totally different results than analyzing each one as though it were an isolated instance.

The most glaring example I can offer is in the draft environmental impact statement prepared for the Air Force for the MX missile system siting in Utah and Nevada. As a member of a task force of the Defense Science Board appointed by the Secretary of Defense to review the draft statement, I observed that the final impact statement was never prepared because the current administration's position on the MX basing system made it irrelevant. The MX system was treated as though it were entering this area without other development of any kind in the western region being considered. That environmental impact statement omitted important elements; it disregarded the fact that energy development requires the same kinds of inputs, requires many of the same kinds of skills as construction of the MX missile system, requires much of the same kind of fabrication facilities, requires going to railheads and unloading things, and requires establishing regional headquarters. The statement seemed to assume that the MX dropped from the sky into Nevada and Utah, and nothing else was happening adjacent to the immediately affected area.

The foregoing example suggests our need to develop our ability to model the entire system with which we are trying to deal. That doesn't mean in every detail, but it does mean that researchers should try to include all the foreseeable, significant developments. Researchers should not consider every individual project in isolation; instead, researchers should consider projects as parts of regional systems that have many components.

In the Southwest study mentioned earlier, we tried to include regional or systems considerations. We deliberately set out to take into account all the significant developments that we could. We developed models that forecast or projected for different assumptions the economic and demographic effects, emissions to the atmosphere, possible water problems, and other effects associated with the entire complex of possible occurrences given different scenarios for development.

Now such efforts are expensive, requiring sustained support for years. The work was participated in at first by the Los Alamos National Laboratory, which continued to support the effort. Much information from the Los Alamos National Laboratory results from further insights gained from the application of these data and models and their further development. We are now at the threshold of improving our ability to do such modeling. Although we have better techniques and are gradually developing data, it looks, at this particular time, as though the entire enterprise may die. Current budget cuts and the trimming of the DOE and the EPA, especially in the research area, probably will force us to proceed in a much greater degree of ignorance in energy development in the West than would otherwise have been the case.

Lest I be regarded as an excessive advocate of modeling, I should caution the reader that modeling is difficult and the phenomenon of looking where the light is that was referred to earlier is real. Recalling the example of the MX missile draft environmental impact statement, let me describe the economic analysis done for the statement. Economists use a technique called input-output analysis. A set of simultaneous equations project, given the technology that is available to the economy, what difference it will make directly and indirectly throughout the economic system if some component of final demand changes. (Final demand here would be exports, consumer demand, or government demand.) That tool was developed for the analysis of relatively small perturbations in a richly interlinked economic system. Of course, one can try to use it under less than ideal conditions.

The specific technique used in the MX environmental impact statement is called Regional Input-Output Modeling System (RIMS). It is a regionally distributed input-output system that was created by the Department of Commerce. It is composed of little input-output systems linked to one another for all the counties in the country. The small systems were produced by applying disaggregation techniques to the national input-output table. Consider counties in Utah and Nevada where Coyote Springs is a big city, where no economic structure exists. Applying this tool to an enormous investment--even the Air Force said it would cost $35 billion--in an area with no economic structure while alleging that it has structure and would behave in the way that a mature economy would behave is bound to produce questionable data. What approach would be better?

There obviously is an area of methodological development for those economists and demographers and related people who concentrate their attention on sparsely populated areas. The tools of modeling created for much more developed areas may be inappropriate. We may have to learn to think in new ways about economic and demographic impacts in such areas.